Photographic Aerial Reconnaissance
and
Interpretation
KOREA, 1950-1952

✈ ✈ ✈ ✈

Yokota Air Base, Japan
Taegu and Kimpo Air Bases, Korea

Photographic Aerial Reconnaissance and Interpretation

KOREA, 1950-1952

✦ ✦ ✦ ✦

Yokota Air Base, Japan
Taegu and Kimpo Air Bases, Korea

by
Ben Hardy and Duane Hall

Sunflower University Press®

1531 Yuma • P.O. Box 1009 • Manhattan, Kansas 66505-1009 USA

Printed in the United States of America on acid-free paper.

ISBN 0-89745-275-5

COVER: Patches: Fifth Air Force and Far East Air Forces
        Campaign Ribbons (left to right): United States Presidential Unit Citation, Korean
                            Presidential Unit Citation
        Top: Air Force Outstanding Unit Award, Army of Occupation Medal, National
            Defense Service Medal
        Bottom: United Nations Service Medal, Korean Service Medal with seven campaign
            stars (1 silver and 2 bronze), Republic of Korea War Service Medal

Sunflower University Press is a wholly-owned subsidiary
of the non-profit 501(c)3 Journal of the West, Inc.

Contents

✈ ✈ ✈ ✈

Dedication

Dedicated To
Colonel Karl L. "Pop" Polifka

✈ ✈ ✈ ✈

Karl Polifka was familiarly known as "Pop" by many of us who had the good fortune to serve under him when he was a Colonel with the 67th Tactical Reconnaissance Wing in Korea. We were deeply grieved by his untimely death on July 1, 1951, just four months after the 67th TRW had been activated. But Pop died doing what he most dearly loved, flying reconnaissance in the line of duty. He was a reconnaissance pilot.

Colonel Polifka started his career in aerial photography on mapping and charting missions in Alaska and Western Canada. Dr. Robert F. Futrell, a military historian who served with both the U.S. Army and the Air Force, wrote in *The United States Air Force in Korea* (1961) that Colonel Polifka was "one of the USAF pioneers in the field of aerial reconnaissance." The Colonel's combat experience began with the 8th Tactical Reconnaissance Squadron (the 15th TRS during the Korean War). He was subsequently transferred to the Mediterranean Allied Photographic Reconnaissance Command and became the outstanding reconnaissance pilot flying in a Lockheed RP-38 Lightning. He later became the unit's Commanding Officer. During the Okinawa campaign of World War II, he had served in the Intelligence section (G-2) as an Air Officer with the Tenth Army.

On February 25, 1951, the 67th Tactical Reconnaissance Wing was activated, and in March units of the 67th were located at Taegu Air Base, South Korea, under Colonel Polifka's command. The 67th was assigned directly to the Fifth Air Force. By pulling scattered units from the Far East Air Forces Command and welding them together, Colonel Polifka soon had a cohesive and vital group that did an excellent job supporting the Allied ground forces.

✈ ✈ ✈ ✈

Acknowledgment

We wish to thank the many friends that have encouraged us and supplied needed advice and direction in the creation of this manuscript, in particular Colonel Karl L. Polifka's son Karl; Sam Dickens, an RF-86 reconnaissance pilot; and R. Cargill Hall, formerly of the National Reconnaissance Office historian, who searched through the National Archives and retrieved photos of Uiju Airfield for us; as well as the very courteous staff at the USAF Historical Research Agency, Maxwell Air Force Base, Montgomery, Alabama.

Many of the photographs herein are from U.S. Air Force aerial reconnaissance missions; one is from the U.S. Navy, and a few are our own. Those supplied by friends have been thus acknowledged.

Duane Hall, Vernal, Utah
Ben Hardy, Santee, California

Introduction

Numerous books and articles have been written on the Korean air war, the foot soldiers, and the tank battles. But there is a dearth of information about the vital technical support provided to combat forces in Korea. Never before in the history of warfare were ground forces so rapidly supplied with photo intelligence as they were in the Korean conflict. The photo interpreters, along with the photo laboratory personnel and the unarmed pilots in the reconnaissance aircraft, did a Herculean job in providing intelligence reports and large-scale photographs of the front-line sectors to the Commanders.

Providing daily reports on the enemy order of battle, along with the status of airfields, transportation lines, artillery and antiaircraft installations, industries, and front-line reconnaissance was an around-the-clock task for the intelligence people. Within minutes after a reconnaissance aircraft had returned to base, the exposed film was rushed to the Reconnaissance Technical Squadron (RTS) for processing. After the prints were made and sent to the Photo Interpretation (PI) Department, they were scanned for vital information to be flashed to the Joint Operations Center (JOC). This book is an attempt to describe the aerial reconnaissance and photo interpretation so vital in support of the combat forces.

Both of the authors had the good fortune to be assigned to the Far East Air Forces (FEAF) when the Korean War started in June 1950. They were members of the 548th Reconnaissance Technical Squadron at Yokota Air Base, Japan. Good luck continued to hold when they were transferred in early 1951 to the 67th Reconnaissance Technical Squadron (redesignated from the 363rd RTS) based at Taegu Air Base, Korea. Later, still, they moved up to Kimpo Air Base, northwest of Seoul.

This manuscript is not the complete story of what transpired with photo intelligence/photo interpretation during the Korean War, but rather the activities that took place while the two authors were stationed with the Far East Air Forces — a very personalized recounting of their experiences during the early stages of the war. Any errors or omissions, or commission, are unintentional. Fifty years can greatly diminish memories, and most of this story is based on the authors' recollections, with a little help from the historical record.

Compared with the equipment currently available, our Korean War efforts in photo intelligence/photo interpretation were primitive — yet highly effective. Despite the advent of robotic drones like the contemporary Predator and the Global Hawk, and integral satellites, it is our belief — perhaps biased — that reconnaissance pilots and their support groups are still an essential part of our military.

Korean place names in this book are from the Army Map Service (AMS) L551 series of charts, and all previously classified documents herein have been declassified by Executive Order (EO) 12959.

✈ ✈ ✈ ✈

Photographic Aerial Reconnaissance and Interpretation
June 1950-February 1952

THE first recorded use of "aerial reconnaissance" in combat by the United States took place during the Civil War. It was in the midst of General George B. McClellan's siege of Richmond, Virginia, and was accomplished from a tethered balloon. Later, during World War I, photo reconnaissance came of age and proved to be a vital and necessary function, though primitive by today's capabilities. Photo interpretation then progressed into full bloom during World War II. The British Royal Air Force (RAF) brought it to the forefront of strategic and operational planning.

When the United States entered the war in December 1941, it soon formed a liaison with the British and set up an Allied Reconnaissance Wing in Tunis, North Africa. Colonel Elliott Roosevelt, President Franklin Roosevelt's second son, was in command of this new unit in 1943, with Lieutenant Colonel Karl L. "Pop" Polifka as his second in command. As things progressed, Polifka became the outstanding reconnaissance pilot and organizer of that wing. Constance Babington-Smith, the famed World War II photographic interpreter working with the Allied Central Interpretation Unit, wrote an excellent book about photo intelligence during World War II. *Air Spy* demonstrated that photo interpretation was an essential factor contributing to the defeat of the Axis powers.

In 2001, during the Afghanistan invasion that followed upon the September 11, 2001, terrorist attacks in America, and subsequently in Operation Iraqi Freedom in 2003, aerial reconnaissance also played a vital role, with U-2 aircraft and unmanned aerial vehicles (UAV) such as the Global Hawk and the Predator.

Most war stories usually are about combat infantrymen, tank crews, air crews, and naval forces, both surface and underwater, as well they should be. These individuals are the ones immediately confronting the enemy and death, and they are the ones who win or lose the battles. (In addition, they bring glamour and glory to a horde of news reporters covering their combat activities and perils.) But non-combatant personnel behind the front line represent a cadre of unsung support people: cooks, bakers, mechanics, medics, supply, communications, military and air police, truck drivers, and intelligence personnel. They enable the front-line troops to continue their task of defeating the enemy.

Sometimes we find these foot soldiers and support cadres (including the U.S. Air Force) unprepared for sudden and unexpected combat, as happened in Korea and Japan in June 1950. The few U.S. military personnel stationed in Korea at that time were assigned to the Korean Military Advisory Group (KMAG) and had no combat mission or capability. The unanticipated explosion on Korea's 38th Parallel caught all of us of the Far East military establishments completely off guard. A couple of days after the invasion of the Republic of Korea (ROK) by the Communist North Korean People's Army (NKPA), President Harry Truman made the decision to intervene, with the United Nations' blessing (thanks to the absence of a piqued USSR at the UN). But American troops that rushed to South Korea from Japan were sorely unprepared as they were shoved into front-line combat on very short notice. They were therefore pushed back by the North Korean People's Army into the Pusan Perimeter (sometimes referred to as the Naktong

Perimeter). Meanwhile, hurried support was being sent from outlying areas of the Far East, the United States, and America's allies within the United Nations.

Along with the combat forces, the intelligence community was also completely unprepared to provide the front-line support aircraft and the intelligence and targeting material needed by the 19th Bombardment Group, which was moved some 1,200 miles from Guam to Kadena Air Base on Okinawa, off the southern tip of Japan. The only targeting materials U.S. forces had were in some obsolete target folders we found in an old filing cabinet. The data had been compiled using some very inadequate photo coverage flown during World War II. The only topographic charts were from the Japanese, which the U.S. Army had acquired at the end of World War II. Though excellent, these charts had not been completely translated into English.

The Far East Air Forces, under Lieutenant General George E. Stratemeyer, along with other Air Force commands, earlier had suffered devastating budget cuts. Air Force funding emphasis was on the Strategic Air Command (SAC) and its Convair B-36 Peacemaker and Boeing B-47 Stratojet intercontinental bomber programs. All that remained in FEAF in an aerial reconnaissance role in early 1950 was a disjointed and disorganized, barely still-functioning group.

At the time, the 548th Reconnaissance Technical Squadron was attached to FEAF Headquarters in Tokyo, but sited about 30 miles west at Yokota Air Base, Japan, for logistics. The 548th RTS had two detachments, one at Kadena Air Base, Okinawa, and the other one at Clark Air Base, Luzon, The Philippines. No photo interpreters were stationed at either of these locations, only photo lab personnel. Also at Yokota Air Base was the 8th Tactical Reconnaissance Squadron, which in March 1951 would be redesignated the 15th RTS, flying the Lockheed RF-80A Shooting Star (RF, Reconnaisance Fighter). The 31st Strategic Reconnaissance Squadron (a SAC unit, which later, in November 1950, would be redesignated as the 91st Strategic Reconnaissance Squadron) flying the Boeing RB-29 Superfortress (RB, Reconnaissance Bomber) was stationed at Kadena Air Base. The 6204th Photo Mapping Flight, with two of the Boeing RB-17 Flying Fortress, was stationed at Clark Air Base. These units were in no way capable of providing the close ground support needed by the United Nations forces struggling to hold Pusan, South Korea's principal port on the southeast tip of the peninsula.

The function of the reconnaissance squadrons was twofold: to obtain aerial reconnaissance photography of the enemy-held territory, and to provide visual reconnaissance for the front-line forces and the day and/or night bombers, spotting enemy positions and vectoring in ground fire or attack aircraft to enemy positions. On completion of the photographic missions, the film was delivered to the Reconnaissance Technical Squadrons, which were charged with processing it and making photographic prints. These prints were then sent to the Photo Interpretation Sections to compile intelligence reports. The reports, along with a set of photos, were sent to the various user organizations. The Reconnaissance Technical Squadron also updated maps from the photography and published them for distribution.

The Photo Intelligence people (and the photo interpreters, who we called the PIs) number among those most frequently overlooked in accounts of the Korean War. Because of the classified nature of their participation, very little has been written about this important group and their intelligence reports and aerial reconnaissance photographs. The ground soldier's war is fought within the range of his rifle or howitzer. The war of the unarmed aerial reconnaissance pilot and the photo interpreter is everywhere on the enemy's side of the front line and sometimes across the border of neighboring countries.

The PIs monitored all enemy activity in the whole of North Korea and the surrounding territory: location and condition of the airfields; aircraft count; antiaircraft and radar sites; number of tanks, trucks, and artillery pieces; ammunition dumps; railroad rolling stock; power plants; industrial plants; rail yards; highway and railroad bridges; and war equipment, including hundreds of tons of supplies — food, clothing, and fuel (termed POL, petroleum, oil, lubricants) — to be targeted and destroyed behind the enemy lines. The destruction of all of these, without a doubt, directly and adversely affected the enemy at the front line. And this activity was all made possible from the information provided by aerial reconnaissance photography.

The photo interpreter gleaned all the information he could from a thorough study of the aerial photos. Frequently we coordinated and verified information obtained from Prisoner of War Special Interrogation Reports. In those days, a PI's equipment was extremely simple: a small stereoscope, which allowed a three-dimensional view of the photos and which folded up and fit in a shirt pocket (we still have ours!); a tube magnifier (about the size of a roll of quarters) marked with a 1/1,000th-foot scale; a ruler with the same scale; and a slide rule for measurement calculations. With these pieces of equipment and proper schooling, which we had, it was amazing how much intelligence information could be gleaned for the Photo Intelligence Report from good, clear, reconnaissance photographs.

Late on Sunday afternoon, June 25, 1950, rumors were floating around Yokota Air Base of reports from the Armed Forces Radio that the Democratic People's Republic of Korea (North) was invading the Republic of Korea (South). No one seemed particularly concerned. Most men didn't even know where Korea was located. But things changed drastically during breakfast on Monday morning with an announcement on the public address system: "ALL PERSONNEL REPORT TO YOUR DUTY STATIONS." The order was followed by the offhand comment from the men, "*What the hell now?*" Shortly a second message came through. We were to report "*ON THE DOUBLE!*" This announcement sounded much more serious, and soon the men started heading for their assigned work stations, though admittedly not "ON THE DOUBLE," for things at Yokota had been slow and peaceful in the occupation of postwar Japan.

On arrival at our duty stations that day, we were informed that the North's Democratic People's Republic had indeed launched the reported invasion of the Republic of Korea. But additional information, as to what was happening across the Korea Strait, the some 110-mile-wide channel between Japan and Korea, was unavailable to the Squadron Commanders. The best source of news, once again, was the Armed Forces Radio station.

The men of the 548th Reconnaissance Technical Squadron, under the command of Major George H. Fisher, were among the groups responding to that early-morning message. Staff Sergeant Ben Hardy and Private First Class Duane Hall were photo interpreters assigned to the Photo Intelligence Department, headed by Captain Adrian M. Burrows, of the 548th RTS. Duane Hall, along with Privates First Class Robert G. Selsor and Philip M. Simons, had arrived at the 548th Reconnaissance Technical Squadron on March 20, 1950, from the 5th Reconnaissance Technical Squadron, Travis Air Force Base, Fairfield, California. Staff Sergeant Hardy, along with Privates First Class John W. Beisel, Carol L. Evans, Marlyn C. Fetter, and Robert A. Johnston, from the 91st Reconnaissance Technical Squadron, Barksdale Air Force Base, Shreveport, Louisiana, had arrived at the 548th RTS on May 20, 1950. Both Hardy and Hall, in 1949, had completed the Photo Interpretation Technician course at the USAF Air Training Command, Lowry Air Force Base, Colorado. Hardy, in 1948, had completed the Air Training Command Photography course.

At our duty stations that Monday, June 26th, we were given a short briefing on maintaining security of classified material and of our work area. We were also informed that the entire base was on an alert status, and tight security was to be maintained in our work area as the squadron was involved in several highly classified projects. There was some discussion that day of the Soviet dictator, Josef Stalin, and his continuing boast about subverting the world to Communism. But we shortly learned that President Harry Truman had made his own decision, "drew a line in the sand," and let it be known that "*We will stop you here!*" At that point the United Nations had what was called a "Police Action" on its hands. The politicians had decided not to refer to it as a war, but rather by the euphemistic term, and suddenly we were all "policemen."

At the onset of the war, Yokota Air Base was a fighter interceptor and photo reconnaissance installation. The 35th Fighter Interceptor Wing (FIW) was composed of the 39th Fighter Interceptor Squadron (FIS), the 40th FIS, and the 41st FIS. All three Fighter Interceptor Squadrons were flying the

Oftentimes the photo interpreters were given Prisoner of War Special Interrogation Reports to check the accuracy of information gleaned from prisoners. This and the photos on the next two pages were enclosures used for verification.

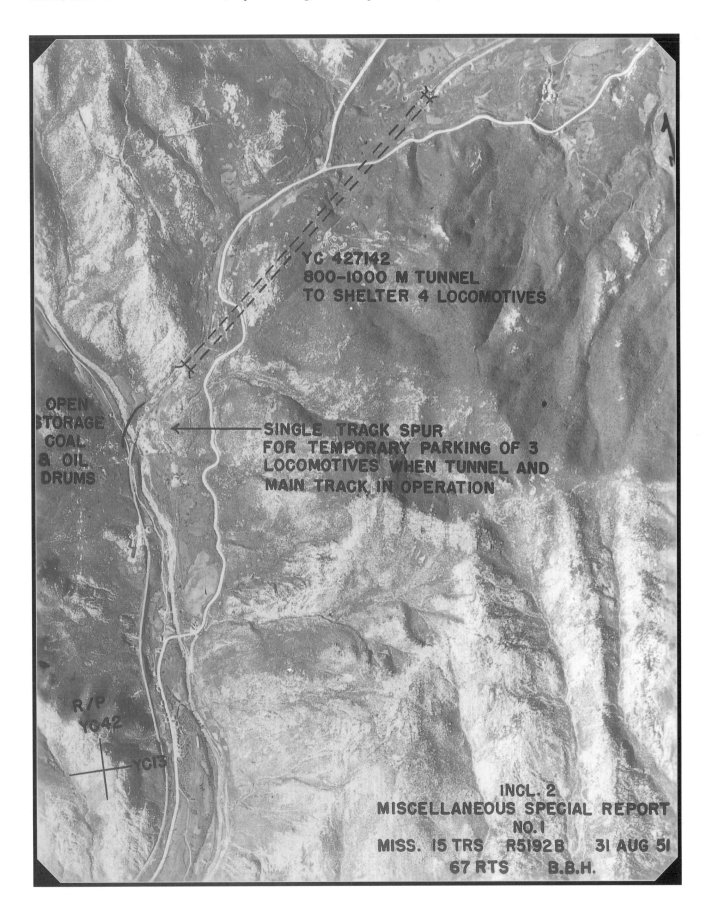

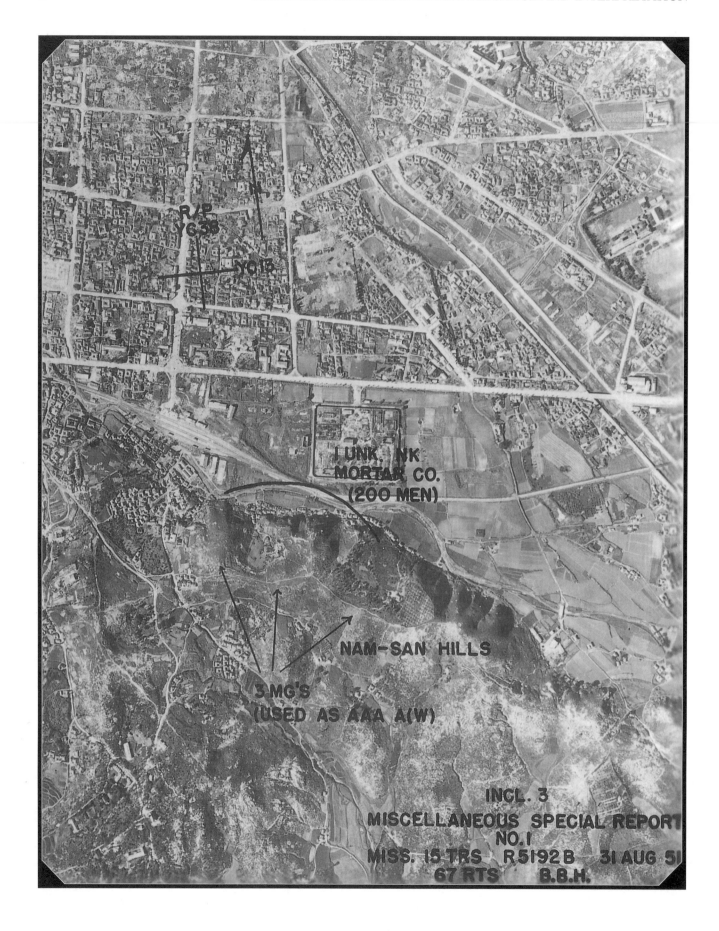

United States Air Force

Air Training Command

Be it known that   SERGEANT BENSON B. HARDY, AF 17227171

has satisfactorily completed the prescribed course of instruction of the

Air Training Command specializing in

PHOTOGRAPHY

In testimony whereof and by virtue of vested authority we do confer upon

him this

═══ CERTIFICATE OF PROFICIENCY ═══

Given at   LOWRY AIR FORCE BASE, DENVER, COLORADO

on this   TWENTY-FIRST   day of   DECEMBER

in the year of our Lord one thousand nine hundred and   FORTY-EIGHT

Attest:

S. T. BUSH
Lt Col, USAF
Supv Photo Tng
SECRETARY

WARREN R. CARTER, Brigadier General, USAF
COMMANDANT

United States Air Force

Air Training Command

Be it known that   STAFF SERGEANT BENSON B. HARDY, AF-17227171

has satisfactorily completed the prescribed course of instruction of the

Air Training Command specializing in

PHOTO INTERPRETATION TECHNICIAN

In testimony whereof and by virtue of vested authority we do confer upon

him this

CERTIFICATE OF PROFICIENCY

Given at   LOWRY AIR FORCE BASE, DENVER, COLORADO

on this   THIRTIETH   day of   AUGUST

in the year of our Lord one thousand nine hundred and FORTY-NINE

Attest:

RICHARD H. CARTWRIGHT
Captain, USAF
SECRETARY

WARREN R. CARTER
Brigadier General, USAF
COMMANDANT

United States Air Force

Air Training Command

Be it known that   PRIVATE FIRST CLASS DUANE HALL, AF-19334602

has satisfactorily completed the prescribed course of instruction of the

Air Training Command specializing in

PHOTO INTERPRETATION TECHNICIAN COURSE

In testimony whereof and by virtue of vested authority we do confer upon

him this

CERTIFICATE OF PROFICIENCY

Given at   LOWRY AIR FORCE BASE, DENVER, COLORADO

on this   FOURTEENTH   day of JUNE

in the year of our Lord one thousand nine hundred and   FORTY-NINE

Attest:

HENRY G. V. HART
Lt Colonel, USAF
Supervisor, DIT

SECRETARY

WARREN R. CARTER
Brigadier General, USAF
Commanding

COMMANDANT

Lockheed F-80C Shooting Star. The 8th Tactical Reconnaissance Squadron, flying RF-80As, and the 339th Fighter Interceptor All-Weather Squadron, flying North American F-82 Twin Mustangs (with their twin engines and twin booms), also were based there. In early July, the 8th Tactical Reconnaissance Squadron transferred to Itazuke Air Base, in southern Japan. Soon after its departure, the 92nd Bombardment Group arrived from the States with its famed B-29 Superfortress bombers, and Yokota Air Base became primarily a Bomber Command Headquarters installation under the command of Major General Emmett "Rosie" O'Donnell. All the B-29 bomber groups in the Far East were placed under his command.

Very shortly after President Truman committed the United States to the defense of South Korea, RB-29s from the 31st Strategic Reconnaissance Squadron based at Kadena Air Base, Okinawa, and RF-80As from Yokota Air Base were winging their way across the Sea of Japan headed for the Korean Peninsula. These RB-29s and RF-80As were fitted with aerial cameras instead of bombs or .50-caliber machine guns. The cameras in the RB-29s were the K-17 trimetrogon; their film magazines usually held a 390-foot roll. The K-17 was a three-camera installation, with a six-inch focal-length lens, which produced 9x9-inch vertical as well as overlapping left and right oblique photos. This gave us horizon-to-horizon coverage on a very small scale. The photos were of primary use in locating targets, updating topographic charts, facilitating the plotting of the large-scale reconnaissance photography, and creating mosaics.

Pairs of K-22 vertical cameras were used as well, which also had film magazines with 390-foot rolls, usually with a 40-inch focal-length lens that gave excellent overlapping 9x18-inch large-scale photos producing stereo vision — well suited for writing intelligence reports. The RF-80As usually mounted two K-22 vertical cameras with either a 24-, 36-, or 40-inch focal-length lens and usually a K-17 forward oblique camera.

In August 1950 the 162nd Tactical Reconnaissance Squadron (Night Photography) arrived from Langley Air Force Base, Virginia, with its Douglas RB-26 Invader tactical light bomber aircraft. Later, in February 1952, the 162nd was redesignated the 12th Tactical Reconnaissance Squadron. The RB-26 aircraft had a pair of K-19 cameras, with magazines holding 200-foot rolls of film that produced 9x9-inch photos. But in its early sorties, the 12th TRS experienced extreme difficulties with a high rate of malfunctions from outdated flash bombs.

The 45th Tactical Reconnaissance Squadron was activated in early September 1950, but did not receive its North American RF-51 Mustangs until November. The cameras in these aircraft were usually K-22s, vertically installed, that yielded 9x18-inch photos.

The many different cameras used in the theater gave us a variety of photographic prints. The best and most commonly received photos were the 9x18-inch verticals that varied in scale from about 1:8,000 to 1:10,000. Oftentimes the forward oblique photos helped to identify objects. The 452nd Bomb Group had a tail-mounted oblique K-25A camera that gave us interesting 5x5-inch pictures of low-altitude bombing results.

After a six- to eight-hour reconnaissance flight, the aircraft returned to Yokota Air Base and the film was rushed to the 548th Reconnaissance Technical Squadron Operations Center where it was prioritized and sent to the photo lab for processing and printing. In the meantime, the pilot's flight plot had gone to the Photo Intelligence Section in order that the correct aeronautical charts could be obtained on which to plot the mission. The negatives were annotated, using a grease pencil (also known as a wax pencil for writing on film, glass, or porcelain-type surfaces, and easily removed), with the exposure number, mission number, and date flown for the preliminary printing. On an initial printing, three sets were made. Two were delivered to the Photo Intelligence Department, one set for intelligence evaluation and the other for plotting the mission, while a third set was delivered to the Photo Mapping Department to be used in revising and updating topographic charts. Upon receipt of the photographs, the photo interpreter did a quick scan of the prints to see if there was any vital information that needed to be called in immediately to the Joint Operations Center. An Immediate Photo Intelligence Report (IPIR) was initiated and submitted for typing, reproduction, and distribution to using agencies. A more detailed Mission Review Intelligence Report (MRIR) was then initiated.

After this report was typed up and mimeographed (no handy photocopy machines in those days), it was joined up with a set of prints of the mission photographs and distributed to the Joint Operations Center, the Headquarters Far East Air Forces, the Far East Command (FEC), the Photo Interpretation Department in the Pentagon, the Eighth Army Command, the Strategic Air Command, and other using agencies. Another assigned task for the 548th Reconnaissance Technical Squadron required the processing of gun-camera film, as the 548th was the only unit with the equipment to process that film in the early stages of the war.

The maps we had to work with were the U.S. Air Force aeronautical charts at a scale of 1:250,000 — not very useful for fine detail, but good for plotting the reconnaissance mission's photography. Fortunately, we had the U.S. Army maps (AMS L551 and L751 Series) acquired from the Japanese army,

which were being updated by the U.S. Army 64th Engineering Topographic Battalion in Tokyo. The Photo Mapping Department of the 548th Reconnaissance Technical Squadron was also updating charts from the reconnaissance missions. These maps were at the scale of 1:50,000 and 1:250,000. The larger-scale maps had excellent details and were great for locating specific sites and installations.

Many of the names on the maps for locations in North Korea were in Japanese. In the process of having these maps translated into English, some Japanese designations were retained, following Japan's occupation of Korea, where they had established cities and industries. In many cases, we did not know the names used by the North Koreans. In checking some of the place names in our reports and maps with a *National Geographic* atlas, we found several variances; for example, Antung,

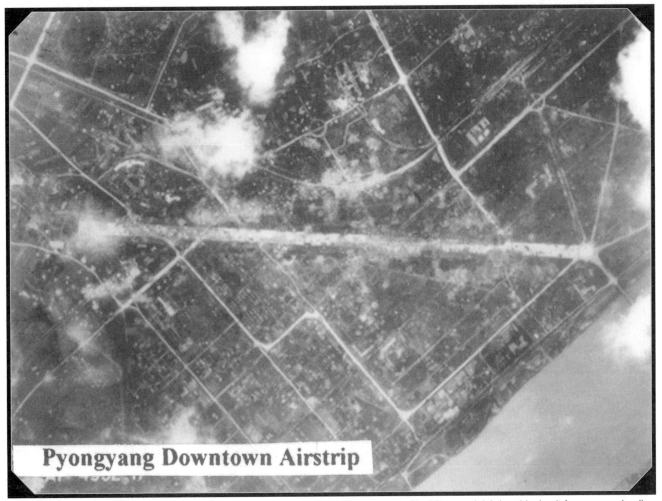

Pyongyang Downtown Airstrip

Pyongyang Downtown Airstrip, a paved city street that the North Koreans had cleared of buildings and debris, widening it for use as a landing strip.

CHOSEN RIKEN METALS CO., CHINNAMPO
38°43'N   125°22'E

The pre-strike reconnaissance photo of Chosen Riken Metals Co., Chinnampo, North Korea.

CHOSEN RIKEN METALS CO., CHINNAMPO
38°43'N 125°22'E

An excellent saturation bombing of the Chosen Riken Metals Co., by the 92nd Bombardment Group out of Yokota Air Base, Japan, August 31, 1950.

CHOSEN RIKEN METALS CO., CHINNAMPO
38°43'N 125°22'E

A post-strike reconnaissance photo showing the results of a bombing attack on the Chosen Riken Metals Co.

The Mitsubishi Light Metals Co., Chinnampo, still another North Korean industry bombed out of existence on August 31, 1950, by the 98th Bombardment Group from Yokota Air Base, Japan.

across the border in Manchuria, is now Dandong, and Chinnampo, North Korea, site of the Chosen Riken Metals Company (a former Japanese plant, and Chosen being an ancient name for Korea), is now Nampo. During 1950-1952, however, we only had the Japanese names for these facilities, not the North Korean, so our reports retained those designations. You will notice that on some of the photographs in this manuscript certain industrial installations have the Japanese nomenclature. A current Korean map (July 2003, *National Geographic*) shows the new spellings: Kimpo is Gimpo; Inchon is Incheon (site of the new International Airport, which replaced the one at Kimpo); Taegu is Daegu; and Pusan is now Busan.

During the conflict, the North Koreans kept building new airfields, particularly in the MiG Alley area next to the Chinese border. After we had located one of those installations and had plotted it on the large-scale map, we would name it after the town it was near. One of our favorite names was the one we assigned as Pyongyang Downtown Airfield. We had noticed some unusual activity on a long, straight, wide street in Pyongyang. The North Koreans had cleared debris and buildings adjacent to that paved city street and widened it to use as a landing strip. They were evidently planning to use cross-streets as taxiways to utilize the aircraft revetments that had been built amidst the nearby buildings. Naturally, we had no way of knowing what name the North Koreans had assigned to this particular airfield.

A further indication of the American unpreparedness at the start of hostilities in June 1950 was the lack of photo interpretation personnel assigned to the 19th Bombardment Group of the Twentieth Air Force stationed at Kadena Air Base, Okinawa. Four men — First Lieutenant Charles Curl, Staff Sergeant Ben Hardy, and Privates First Class Jack Kucker and Bob Selsor — were flown down from the 548th Reconnaissance Technical Squadron on temporary duty (TDY) to support the bombardment group. They were there to plot the bombing missions and write bomb-damage and assessment reports from the film brought back by the B-29 bomber crews. It was in an old filing cabinet in the 548th Reconnaissance Technical Squadron Detachment 1 photo lab, in which the old target folders previously mentioned were found. We used the as-

signed target numbers from these folders when writing our Mission Review Intelligence Reports. Apparently these folders had been compiled by the 6th Photo Technical Squadron (forerunner of the 548th RTS) when it became obvious that the USSR was not going to allow the Korean Peninsula to become united.

A query from Far East Air Forces Headquarters asking us about the source of these target numbers left us baffled. FEAF had no record of those target folders and target numbers, and we were directed to refrain from using the numbers, as they had no valid meaning or status. However, the bomber crews continued to refer to the photos from these folders for target identification. In addition, we had requested the bomber crews to keep their cameras operating while flying over enemy territory to help us detect and locate any new military installations or activity that would assist in future mission targeting.

After his return to the 548th Reconnaissance Technical Squadron from Okinawa on August 19, 1950, Staff Sergeant Ben Hardy was assigned to Captain Byron Schatzley's team, working on the project interpreting the oblique 9x18-inch photographs taken by the RB-29 with its K-30 100-inch focal-length camera. These missions had been flown along the coastal areas of China, Siberia, and the Russian-occupied Kuril Islands between the Russian Kamchatka Peninsula and northern Japan — missions that had commenced prior to the Korean War. During this same period, Sergeant Duane Hall was on temporary duty in the photo lab, working on the highly classified "Wringer" project, the 1949-1955 effort at "containing" the Soviet Union. The 548th's contribution to Wringer was the photocopying of some 200,000 Japanese government service records of World War II Japanese military personnel that had been repatriated back to Japan by China.

Two critical and outstanding reports were compiled by the 548th Reconnaissance Technical Squadron during July and August 1950. General of the Army Douglas MacArthur had requested a detailed study of the Inchon Harbor area, the port city for Seoul, the capital of the Republic of Korea. This produced *Terrain Study #13, Seoul and Environs,* an in-depth report in preparation for what would be the September 15, 1950, invasion at Inchon Harbor. *Study #13* led to the successful (and widely acclaimed) surprise landing that provided a quick sweep into Seoul, which helped break the back of the North Korean invasion forces attacking the Pusan Perimeter and trapped a large force of enemy troops in the southern portion of the peninsula. The 548th RTS received many letters of commendation: September 13, 1950, and February 1, 1951, from USAF Lieutenant General George E. Stratemeyer,

---

THE COMMANDING GENERAL
FAR EAST AIR FORCES
APO 925

13 September 1950

SUBJECT: Commendation

TO : Commanding Officer
   548th Reconnaissance Technical Squadron
   APO 328

THRU : Commanding Officer
   Far East Air Forces Base
   APO 925

  1. I have followed, with great satisfaction, the work that is being accomplished by your organization both as to the quantity and technical quality of your product. I was especially interested in your Production Report for the month of August in which you indicate that you have produced over two hundred thousand photographic prints and two and a half million impressions in your printing plant.

  2. I commend you, and through you the officers, airmen and civilians of your command, for an outstanding accomplishment in the present emergency. Your valuable contribution to the Korean War effort is most gratifying and greatly appreciated. Keep up the good work.

GEORGE E. STRATEMEYER
Lieutenant General, U. S. Air Force
Commanding

BASIC: Ltr fr CG, FEAF, APO 925, subject, "Commendation", dtd 13
       Sep 50.

200.6   1st Ind

Headquarters, Far East Air Forces Base, APO 925, 15 September 1950.

TO: Commanding Officer, 548th Reconnaissance Technical Squadron, APO
     328.

     Forwarded with deep pride and a reiteration of my pledge to
assist you in every possible way.

                                   MAURICE M. SIMONS
                                   Colonel, USAF
                                   Commanding

                               2nd Ind

HEADQUARTERS, 548TH RECONNAISSANCE TECHNICAL SQUADRON, APO 328   18 Sep 50

TO:: Officers and Airmen, 548th Reconnaissance Technical Squadron, APO 328

     1. I wish to take this opportunity to express my appreciation to the
officers and airmen of this organization for their untiring endeavor and
accomplishment of a difficult and arduous duty during this time of emergency.

     2. Your attention to duty and willingness to work long hours to get
the job done has been very gratifying.

     3. The cooperative spirit of all concerned is indicative of the
high morale and team-work which is so desirable in the military service.

                                   GEORGE H. FISHER
                                   Major, USAF
                                   Commanding

expressing his satisfaction with the organization's work; September 13, 1950, from Colonel Maurice M. Simons, USAF; September 18, 1950, and February 1, 1951, from Major George H. Fisher, USAF; and October 25, 1950, from Colonel Claude E. Putnam, 92nd Bomb Group. On October 10, 1950, C. B. Cates, of the U.S. Marine Corps Commandant's Office, and on November 1, 1950, U.S. Army Major General C. A. Willoughby also commended the 548th's work on the terrain and defense study for the Inchon landing, code-named Operation Bluehearts. Subsequently received was a letter of January 15, 1951, from Colonel O. H. Rigley, Jr., noting the "excellent" reports, and a letter of appreciation dated March 1, 1951, from USAF Lieutenant Colonel Richard L. Walker, Chief of the Far East Air Forces Project Wringer.

The 548th's second report had been a detailed study and site selection for the October 20, 1950, parachute assault by the 187th Airborne Regimental Combat Team at Sukchon and Sunchon, north of Pyongyang, North Korea.

✈ ✈ ✈ ✈

HEADQUARTERS U. S. MARINE CORPS

COMMANDANT'S OFFICE

WASHINGTON

10 October 1950

Dear General Willoughby:

We arrived back home yesterday morning after a most interesting and enlightening trip. I was impressed with what I saw while in the Far East and would like to particularly commend you for the terrain and beach study which you prepared for the Inchon landing.

With best wishes and kind personal regards,

Very sincerely,

C. B. CATES

HEADQUARTERS 92D BOMBARDMENT GROUP (M)
APO 328

25 October 1950

SUBJECT: Letter of Commendation

THRU: Brigadier General C.Y. Banfill
           Deputy for Intelligence, FEAF
           APO 925

TO: Major George E. Fisher
           548th Reconnaissance Tech Squadron
           APO 328

    1. It is with pleasure that I commend you and all members of your
command, who have as your recorded accomplishments indicate, performed
magnificently in support of the 92d Bombardment Group (M) during the
period from 8 July 1950 to 25 October 1950. Careful analysis reveals
that many and varied problems confronted you and your organization. You
were capable of resolving these difficulties in a superior manner, thus
realizing maximum utilization of facilities in your control. As a re-
sult of your unremitting, prompt and continuous efforts you were always
able to meet this Group's photographic requirements and in most cases
with greater speed than anticipated. You thereby contributed greatly
in the capability of this Group to continually launch maximum efforts
against the enemy. The countless hours devoted to the arduous task of
supplying strike photographs, target materials and allied photographic
needs to this Group and the willingness of you and your personnel to
accept the added responsibilities imposed is both noteworthy and commendable.

    2. You and each member of your organization has my personal thanks,
as well as my official commendation for the contribution the 548th reconnaissance Technical Squadron has made in supporting this Group in
battle.

                                     CLAUDE E. PUTNAM
                                     Colonel,   USAF
                                     Commanding

Though the 548th Reconnaissance Technical Squadron was well established with sufficiently trained personnel and modern equipment, logistics was a primary problem. It soon became obvious that the distance and time consumed by the reconnaissance aircraft flying from Japan to Korea, then returning to Japan, proved to be inadequate to support the UN combat forces. Processing the film, writing the Mission Review Intelligence Reports, and getting the reports and photographs back across the Sea of Japan produced delays that were intolerable to the Eighth Army. And weather similarly could be a delaying factor.

A more versatile and immediate response was mandatory if the ground forces were to operate effectively. With mounting complaints from the Eighth Army, and after an unconscionable delay, the U.S. Air Force finally responded. The limited range of the RF-80, flying from Yokota Air Base, also hindered the coverage of North Korea.

As an attempt to rectify part of the problem, in early July 1950, the 8th Tactical Reconnaissance Squadron had been transferred to southern Japan's Itazuke Air Base, which somewhat cut down the flying distance to the targeted areas in Korea. But this was still not a satisfactory answer. After the aircraft returned from its mission, the film then had to be flown up to the 548th Reconnaissance Technical Squadron for processing and interpretation. And the fall in Japan is typhoon season, with its inherent weather delays.

In a further attempt to provide a timely delivery of the photo interpretation reports and reconnaissance photographs, the 363rd Reconnaissance Technical Squadron had been moved from Langley Air Force Base, Virginia, under the command of

Major General Charles A. Willoughby, USA
Assistant Chief of Staff G-2
General Headquarters, FEC and SCAP
APO #500
San Francisco, California

1st Ind

G-2, GENERAL HEADQUARTERS, FAR EAST COMMAND, APO 500,
    1 November 1950.

TO: Lt Col C. J. Long III, Director of Reconnaisance
     Deputy for Intelligence, FEAF, APO 925

Above commendation for Terrain Study #13, Seoul and Environs, is forwarded with appreciation for the production of a large number of copies of air photographs in a period of two days. Inclusion of original photographs materially increased the usefulness of the volume and contributed to its success.

C. A. WILLOUGHBY
Major General, GSC
Assistant Chief of Staff, G-2

C O P Y   **RESTRICTED**   C O P Y

DEPARTMENT OF THE AIR FORCE
HEADQUARTERS UNITED STATES AIR FORCE
WASHINGTON 25, D. C.

AFOIN-C/RC   15 Jan 1951

SUBJECT: (Unclassified) Special Photo Intelligence Reports

TO   : Commanding General
           Far East Air Forces
           Tokyo, Japan
           Attn: Director of Intelligence

    1. The Special Photo Intelligence Reports issued by the 548th
Reconnaissance Squadron, FEAF, have been excellent in quality and of
great value in the production of further intelligence by Headquarters
USAF.

    2. Strategic Air Command has a requirement for these Special Photo
Intelligence Reports and it is requested that they be included on the
distribution list for four (4) copies, enabling them to make further
distribution of one (1) copy each to the Second, Eighth, and Fifteenth
Air Forces. If sufficient copies of the previously issued Special Photo
Intelligence Reports are available, it is further requested that four
(4) copies each of these be forwarded to Strategic Air Command. This
would in no way alter the presently operating regular distribution of
the Mission Review Photo Intelligence Reports.

        BY COMMAND OF THE CHIEF OF STAFF:

                                /s/ O. H. Rigley, Jr.
                                /t/ O. H. RIGLEY, JR.
                                    Colonel, USAF
                                    Executive, Collection Div.
                                    Directorate of Intelligence

RESTRICTED

Major Floyd W. Brewer, to Itazuke Air Base late in August 1950, along with the 162nd Tactical Reconnaissance Squadron (Night Photography). To fill the void created by the move of the 363rd RTS, the 67th RTS was activated at Langley Air Force Base with a skeletal group of personnel. The 363rd and 162nd were placed under the command of the newly formed 543rd Tactical Support Group (TSG). In October, these groups moved to Taegu Air Base, just southeast of the city of Taegu, in South Korea, after the Inchon landing had removed the pressure on the Pusan Perimeter and sent the North Koreans retreating northward. (Taegu Air Base had been code-named K-2, when FEAF assigned "K" designations to the airfields in Korea,

both North and South.) But without satisfactory facilities to process the large backlog of film at the air base, the 363rd RTS was moved to a school building in the city of Taegu. The disorganized and unsettled conditions continued, with the eventual final and complete loss of the photo lab, all of its equipment, and all reconnaissance film from the 12th TRS and 45th TRS in a fire on March 3, 1950. The 363rd then moved back to Taegu Air Base.

With a priority "rescue mission," much of the equipment lost in the fire was replaced from units in Japan. But in order to process the aerial film and photos, a large supply of clean water was a mandatory requirement and a constant source problem. With a "jerry-rigged" operation of two 3,000-gallon

---

THE COMMANDING GENERAL
FAR EAST AIR FORCES
APO 925

1 February 1951

SUBJECT: Letter of Appreciation

TO   : Commanding Officer
         548th Reconnaissance Technical Squadron
         APO 328

    The attached letter from the Directorate of Intelligence, United States Air Force, with indication of a job well done, affords me a great deal of pleasure. I want to add my personal appreciation to you and, through you, to the members of your organization who both individually and collectively have elicited the favorable comments, and who will, I am certain, continue to merit it by maintaining the same high standards of accomplishment in their future tasks.

GEORGE E. STRATEMEYER,
Lieutenant General, U. S. Air Force
Commanding

Incl:
    Cy of USAF ltr

tarpaulin water tanks and salvaged piping from the burned-out lab, running water in the photo lab was back in place. In addition, a 6,000-gallon water tanker was assigned to the lab; the tanker was constantly in operation, looking for usable quality water. Meanwhile, however, waste water from the processing had to be carried out to another tanker to be hauled away and disposed of. A supply of ice to cool the photographic processing chemicals was another recurring problem. Yet, within a remarkably short time — three days — the squadron was back to full operation, though still under adverse photo lab conditions and with a severe shortage of trained and experienced photo interpretation and lab personnel; and the critical problem of timing also continued.

To get things completely organized and operational, it was then that Lieutenant General George E. Stratemeyer requested that Colonel Karl L. Polifka, of World War II renown, be assigned to Far East Air Forces to organize a functional reconnaissance wing. On February 25, 1951, at Itazuke Air Base, the 67th Tactical Reconnaissance Wing (TRW) was formed on paper and assigned to the Fifth Air Force. Colonel Polifka was named the Wing Commander. A short time later, on the 21st of March, the Wing was assigned to Taegu Air Base, a short distance inside the former Pusan Perimeter. Several scattered units in the Far East Air Forces and from the States were pulled together and incorporated into the 67th Tactical Reconnaissance Wing. The 543rd Tactical Support Group from Itazuke (Japan) Air Base was renamed the 67th Tactical Support Group (TSG), the 8th Tactical Reconnaissance Squadron was renamed the 15th Tactical Reconnaissance Squadron, and the 162nd Tactical Reconnaissance Squadron was redesignated the 12th TRS. The recently activated 45th Tactical Reconnaissance Squadron retained its original designation. The 363rd Reconnaissance Technical Squadron was renamed the 67th RTS.

After a short period of transition and training, a formidable organization finally emerged that could start to provide the intelligence information and photo coverage of the front-line areas, so critically in demand by the United Nations ground forces.

The 67th Tactical Reconnaissance Wing's squadrons flying the photographic reconnaissance missions during 1951-1952 were the 12th Tactical Reconnaissance Squadron flying RB-26 Invaders, the 15th Tactical Reconnaissance Squadron flying

---

BASIC: Ltr of Appreciation fr Lt. Gen. Stratemeyer to CO 548th RTS dated 1 February 1951

1st Ind

COMMANDING OFFICER, 548TH RECONNAISSANCE TECHNICAL SQUADRON, APO 328   4 Feb 51

TO: Officers and Airmen, 548th Reconnaissance Technical Squadron, APO 328

1.   I wish to add my sincere appreciations to that of General Stratemeyer.

2.   It is a pleasure and a priviledge to command a unit in which the assigned personnel constantly strive for the highest standards of achievement.

GEORGE H. FISHER
Major, USAF
Commanding

HEADQUARTERS
FAR EAST AIR FORCES
APO 925

AG 101.2 (1 MAR 51)   1 March 1951

SUBJECT: Letter of Appreciation

THRU   :   Commanding General
           Far East Air Forces
           APO 925

TO   :   Commanding Officer
           548th Photo Recon Tech Sqdn
           APO 328

　　1.　I would like to take this opportunity to express my appreciation to the 548th Photo Reconnaissance Technical Squadron for the outstanding job they have performed for "Wringer".

　　2.　In the past six (6) months, they have accomplished many voluminous and complicated assignments of photo reproduction at our request, and every job was done promptly, efficiently and in a commendable professional manner. The most recent task that they have undertaken at our request is the photo reproduction of the entire Japanese Demobilization Bureau file of repatriate cards. This job alone entails copying some 350,000 documents. They are currently carrying out this mission in their traditionally fine manner.

　　3.　I am sure that this organization is a source of great pride to you, as it certainly reflects the finest training, discipline and supervision. In my estimation, the Director of Reconnaissance also deserves a compliment for establishing and monitoring the close operational coordination that exists between the 548th Photo Reconnaissance Technical Squadron and the using agencies.

                                        RICHARD L. WALKER
                                        Lt. Colonel, USAF
                                        Chief
                                        FEAF Project Wringer

A Lockheed RF-80 Shooting Star reconnaissance fighter on a mission over North Korea.

RF-80A Shooting Stars (and later North American RF-86 Sabre Jets), and the 45th Tactical Reconnaissance Squadron flying RF-51 Mustangs (later RF-80s).

The 12th Tactical Reconnaissance Squadron covered nighttime enemy activities, providing photo coverage using flash bombs, spotting enemy activities and directing the nighttime flying bombers to targets. The 15th TRS covered daytime enemy-held territory — primarily airfields, rail facilities, highways, and industries. The 45th TRS supported the U.S. Army front line with both visual and photo coverage. This overall assist was a great boon to the ground force Commanders in their engagement with the North Koreans and later the Chinese. Never before in the history of warfare had such vital reconnaissance supported the ground forces to the extent provided during the Korean War.

The Commanding Officer of the 67th Reconnaissance Technical Squadron in April 1951 was Lieutenant Colonel Joseph F. Condon, recently arrived from the 91st Reconnaissance Technical Squadron, Barksdale Air Force Base, in Shreveport,

Louisiana. At first the 67th RTS was sadly understaffed with trained and experienced personnel. Very soon, however, it began drawing volunteers from the experienced 548th RTS, which was well staffed. Besides photo lab personnel, six photo interpreters from the 548th (Second Lieutenant Ben Hardy; Sergeants Duane Hall, Dave Hitson, Dan Mason, and Robert Milota; and Corporal Merv Taylor) gladly filled the vacancies. More than welcome, they all had extensive reporting experience with the reconnaissance photography flown since the start of the war and had an excellent working knowledge of the status of North Korea.

When Lieutenant Hardy reported in, he was a familiar face to Squadron Commander Condon, then a Major. Condon had promoted Hardy to Staff Sergeant when they were both assigned to the 91st Reconnaissance Technical Squadron at Barksdale Air Force Base. After Lieutenant Hardy had saluted, Lieutenant Colonel Condon looked up at him with a smile on his face and queried, "Where in the hell have you been?"

It was a great welcome into a new organization. After reminiscing for a few minutes, Lieutenant

Lieutenant Colonel Joseph F. Condon, Commander of the 67th Reconnaissance Technical Squadron, in September 1951 at his "FIGMO" party upon his transfer to FEAF. (FIGMO, for those uninitiated civilians, politely means, *"Finally I Got My Orders."* The officers to the right of Condon are Colonel Vincent Howard, 67th TRW Commander, and Colonel Bert Smiley, 67th TRW Deputy Commander.

Colonel Condon informed Hardy that there was an urgent need for a Special Airfield Section in the Photo Interpretation Department. Knowing from previous experience Hardy's ambition and work ethic, Condon recommended him to Major Arthur A. Grumbine, Officer in Charge of the Photo Interpretation Department, to set up the new project. Major Grumbine heeded Lieutenant Colonel Condon's recommendation and instructed Hardy to follow through with the project and to choose whichever airman he might need to help him. Major Grumbine was aware of Lieutenant Hardy's experience with reconnaissance photography and the usual types of requests for intelligence information, and thus the framework of this new section would be at Hardy's discretion.

After settling into his new duty assignment, Lieutenant Hardy began to formulate in his mind exactly what he wanted incorporated into each target folder for every airfield in North Korea: aircraft count, if any; length and width of runways and taxiways; defense positions of the airfield; identification of radar-controlled antiaircraft weapons; number of aircraft revetments and where they were dispersed; and the possibilities of expansion — everything anyone could possibly want to know about any airfield in North Korea and adjoining countries. Access to the Far East Air Forces Intelli-

gence Summary, classified Top Secret, would also be needed in their work, which somewhat limited the pool of enlisted airmen who could qualify to participate in the ambitious project.

After Lieutenant Hardy had studied his new assignment, he talked with Sergeant Duane Hall, discussing the details of his new challenge and Hall's possible interest in helping set up the files. Both men had been cleared for Top Secret material while in the 548th Reconnaissance Technical Squadron, therefore satisfying that requirement. However, both were still urgently needed to write Immediate Photo Intelligence Reports and Mission Review Intelligence Reports, with the continuing pressure of the constant flow of new aerial reconnaissance missions and the critical shortage of experienced photo interpreters. Nevertheless, the spare moments that occasionally developed gave them the opportunity to work on the airfield target folders. And as they wrote the IPIRs and MRIRs on all types of targets, they would also make notes of critical intelligence information to be included in the continuing new target folder program.

After determining the format that they would use to build a Target Condition Folder for each North Korean airfield, and those in Manchuria, from the coverage that was available, the project got underway. Anyone needing information concerning any particular airfield would find all the information compiled in one folder. Within the folder was included the most recent photo reconnaissance coverage, Immediate Photo Intelligence Reports, Mission Review Intelligence Reports, a Special Airfield Report that covered the particular field, and any other pertinent information available.

As the war continued, in August of 1950 General Stratemeyer ordered the Fifth Air Force to destroy all primary transportation targets between the 37th and 38th Parallels. Operation Interdiction was thus launched by the Fifth Air Force with the

[handwritten signature]

SECURITY INFORMATION

* * * I M M E D I A T E * * *

67TH RECONNAISSANCE TECHNICAL SQUADRON
67TH TACTICAL RECONNAISSANCE GROUP
APO 970

28 February 1952.

IMMEDIATE PHOTO INTELLIGENCE REPORT NO 800

MISSION NO 15TRS R 8475   ALTITUDE 21,000'

DATE FLOWN 28 Feb 52   FOCAL LENGTH 36"

TOT   0830 (I)

TYPE OF PHOTO VV   QUALITY OF PHOTO   POOR

AYD 4521 BCS PYONGYANG E. A/F, 90% cover, prints 3-6

   No A/C noted. R/W is U/S for its entire length. No activity noted.

AYD 4023 BCS PYONGYANG Middle A/F, 100% cover, prints 9-13

   No A/C noted. Concrete R/W is unserviceable. Approx 4000' of sod landing
area is serviceable. Moderate activity on road thru N.E. dispersal area.

AYD 4021 BCS PYONGYANG D.T. A/F, 60% cover, prints 9-14

   No A/C noted. R/W is unserviceable for its entire length. No activity
noted.

OPPTY CYD 4020 BCS PYONGYANG E. M/Y, 100% cover, print 15

   Two (2) locomotives in steam. One (1) has approx twenty-two (22) cars
attached. Approx thirty (30) serviceable, usable units rolling stock.
Three (3) serviceable thru lines. Flashed to JOC 1330 28 Feb 52 .

D4/jal
67RTS

* * * I M M E D I A T E * * *

SECURITY INFORMATION

An Immediate Photo Intelligence Report.

Above, right, and below: Overall views of the 67th Reconnaissance Technical Squadron's facilities at Taegu Air Base.

Right: The 67th Reconnaissance Technical Squadron's new barracks at Taegu Air Base, soon to be evacuated for a move north to Kimpo Air Base.

"Kitchen" facilities of the 67th Reconnaissance Technical Squadron at Taegu Air Base, 1951.

The 67th RTS "dining hall," Taegu Air Base, 1951.

The 67th RTS "scullery" at Taegu Air Base, 1951, where Sergeant Duane Hall had just washed his mess kit. On the left of the photo is the rear of the water tanker that had supplied the "running" water.

Above: Our living quarters at Kimpo Air Base after a heavy downpour.

Left: The water tank that supplied the photo lab at our new Kimpo Air Base facility.

Top left, left, and above: The town of Taegu, South Korea, home of the 67th Reconnaissance Technical Squadron at Taegu Air Base, 1951.

Below: Lieutenant Ben Hardy leaning on a wiped-out, camouflaged North Korean tank at the perimeter of the Taegu Air Base, summer 1951.

Above: Tomb markers along the side of the road between the town of Taegu and the Taegu Air Base, 1951.

purpose of eliminating all highway bridges, railroad lines, rail bridges, and marshaling yards critical to supplying the Communist troops. However, the program almost reached the point of futility when the Communists used enslaved labor for the repair work.

Target folders for these enemy assets were also included in our target file project and were stored in

two wooden boxes approximately two feet square by one foot deep, both equipped with locks. Readily available were phosphorous incendiary devices to be used to destroy the folders in the event we had to evacuate quickly and were not able to take our valuable files with us. Fortunately, we never had to use the incendiaries. The files eventually became a critical reference source for the Photo Intelligence Department.

Making the transition from the reasonably modern and well-established film and lab facilities, comfortable barracks, and well-stocked post exchange at Yokota Air Base to the forward field facilities at Taegu Air Base could well have been considered "culture shock" — though admittedly our facilities were luxurious compared with those of the front-line troops. Except for the orderly room, we had tents with dirt floors. Our helmet served as a "bathtub," and toilet facilities were the old-fashioned country "two-holers." A combat helmet welded onto a two-inch pipe and "planted" in the ground served as a urinal, and these were scattered around the perimeter of the camp area. Even the Post Exchange was accommodated in a tent. Noisy portable generators provided any needed

Left and above: USO Show at Taegu Air Base, 1951, with Marilyn Maxwell and Bob Hope.

electricity. But we did have one small luxury: when available, a can of Schlitz beer cost only five cents.

Taegu Air Base in August and early September 1951 was not far from the front line, where total mayhem reigned. Occasionally the rumble of artillery fire could be heard, and soon an RF-51 Mustang would leave the airstrip with its "armament" of cameras, to get photo coverage and visual observation of any activity in the area. After the film was processed and when photographs were printed, the PI quickly scanned the prints in an attempt to evaluate the scale of the enemy assault. It was then that he calculated, as near as he could, the approximate number of enemy soldiers, tanks, trucks, artillery pieces, and support equipment. At this time he did not do an in-depth study of the area; that came later. The urgent information he compiled was transmitted quickly to the Joint Operations Center where it was used to plot necessary action and issue strike orders.

After a strike aircraft had completed its assigned mission, another reconnaissance aircraft was dispatched to photograph the strike zone, to note any visual activity the pilot could determine, and to bring back the film for a damage-assessment report. The PI person then could sit back and feel a little inner satisfaction that his reporting may have made a difference. Though he did not have to pull a trigger or throw a hand grenade, the information the PI provided hopefully left a lot of enemy troops and their equipment no longer in a serviceable condition.

A PI's job took on an awesome responsibility at times. When the upper command decided to launch an assault, they needed every scrap of information available: type and number of aircraft in striking distance; enemy troops, tanks, artillery units, and support equipment; where they all were located; and flying hazards, such as the heights of buildings, power lines, radar sites, and antiaircraft gun positions, along with terrain elevations.

As time passed in the spring and early summer of 1951, enemy activity around Taegu Air Base had started to settle down somewhat and new accommodations and facilities were built for our work area and housing — buildings with cement floors!

But most important, we had new latrine facilities with *hot* running water and no more 36-gallon water-storage Lister bags, suspended on a tripod or hung from a tree, and fitted with spigots, full of unappealing, warm drinking water. Naturally, all too soon after we were comfortably settled into our new quarters, the word came down to start packing our squadron equipment and our personal gear. We were moving north!

On the 15th of August 1951 we began the task of getting things organized, packed, and loaded for our trek to Kimpo Air Base (K-14). After we loaded the freight train and a 50-truck convoy and sent them on their way north with our entire 67th Wing and personal equipment, we were flown to Kimpo Air Base, 22 miles northwest of Seoul. Under chaotic conditions, with a visit from typhoon Marge in the midst of our trek north, the lack of proper liaison from Combat Cargo, and the lack of unloading provision at Kimpo and the Seoul railhead, we finally did get all personnel and equipment to our area at the air base.

As a footnote to history, typhoon Marge did in one day what the Fifth Air Force and Bomber Command had been trying to do for weeks. In their interdiction program to cut off supplies to the Communist forces at the front line, the Fifth Air Force and Bomber Command were having little luck with the large enslaved labor force the Communists had for repairing railway and highway bridges. In one day, the deluge from typhoon Marge wiped out all the bypass and repaired bridges in western North Korea transporting supplies and personnel from China.

By August 25 we had completed our move, and with all equipment and personnel in place, we were back to functioning full time. Eventually all supporting units of the 67th Tactical Reconnaissance Wing were pulled together at one location. But the first few days were a nightmare; we had an inadequate number of sleeping-quarter tents, and our new facility was incomplete, which led to putting the lab back into tents.

It was not very settling to know that we were now only 28 miles from enemy-held territory. At the top of our "TO DO" list was the digging of trenches, before another drenching typhoon passed over us or the frigid Siberian weather set in, creating a Frozen Chosen of our area.

Although we were back to living in tents, they were a vast improvement over our early Taegu Air Base quarters. The Kimpo tents, thankfully, had wooden decks for a floor, raised up off the earth. For the winter heating system we had a sand box situated in the middle of the floor with a small fuel-oil-burning stove sitting in it. A 55-gallon oil drum alongside the tent, with a rubber fuel line under the decking, was connected to the stove, though with the below-zero temperatures, the fuel line would occasionally freeze. The solution to that problem was a heated brick, which we kept on the stove. Someone would take the brick and run it along the fuel line tubing to thaw out the moisture in the oil to get it flowing again. (A heated brick wrapped in newspaper and a bath towel placed in the foot of your bed also helped to keep your feet warm!)

Our work area was a new and spacious building, though the mess facilities had reverted back to tents for awhile, as construction continued. Soon after settling in, a new Squadron Commander arrived. Lieutenant Colonel Joe Condon had been transferred to FEAF Headquarters in Tokyo, and Major Schuyler S. Harris of the FEAF reconnaissance branch was his relief.

The Photo Interpretation and Lab Processing Departments were greatly enlarged at our new Kimpo Air Base facility. The Eighth Army Photo Group joined us there, as did five officers and two enlisted men from the Royal Air Force. Our previously understaffed squadron was greatly enhanced with new officer and enlisted graduates from the U.S. Air Force School of Photo Interpretation and the School of Photography, both at Lowry Air Force Base in Denver, where years earlier Duane Hall and Ben Hardy had received their own training.

If efficiency means anything, the 67th Reconnaissance Technical Squadron wrote the book. Maybe the 67th RTS can take some credit for being the original "same-day photo service." A roll of 9½-inch by 390-feet aerial film delivered in the morning, with a request for prints the same day, became a common occurrence within the new facility. Had you made that request before the new set-up at Kimpo Air Base, you might have been considered a candidate for the "funny farm." But by late September of 1951, the photo lab had produced 47,959 prints in one day! In a five-month period, 2,177,000 photos and thousands of intelligence reports had

Major Schuyler S. Harris replaced Lieutenant Colonel Joseph F. Condon as Commanding Officer of the 67th Reconnaissance Technical Squadron in the late fall of 1951.

been provided. The 67th was able to do this with the new expanded facilities, the increased support from the Eighth Army Engineer Photography Reproduction and Distribution (EPRD) unit, and the newly arrived Air Force men from the Photo Interpretation and Photography schools. The photo lab also had received newer reproduction equipment and running water — no more rushing around looking for usable water.

Many of the photos and reports were passed on to the new Eighth Army EPRD, front-line units, Fifth Air Force Headquarters in Seoul, Far East Air Forces Headquarters in Tokyo, the Photo Interpretation unit at the Pentagon, Strategic Air Command units, and other using organizations. Credit should be lavished on the photo lab crews. They had round-the-clock shifts, and they worked in very high-humidity summer heat as well as damp, freezing, winter cold, all without an excess amount of bitching! Their volume of work simply kept pouring in, to meet the requests and requirements of the front-line Commanders and other military units.

Occasionally word would come down to the 67th RTS to "clean up" their work stations and put on khaki uniforms, instead of their usual fatigues, because VIP visitors were arriving (though sometimes the men did not receive advance notice). The first and most prominent of the distinguished visitors at the 67th Reconnaissance Technical Squadron was Secretary of the Air Force Thomas K. Finletter, on June 13, 1951. This visit gave the men of the 67th RTS the opportunity to brief him on their intelligence reports and activities. The next arrival was the Chairman of the Joint Chiefs of Staff, General

The cover of the 67th RTS 1951 *Annual.* In the center of the photograph is the 67th's work facility and the Eighth Army Photo Group, at Kimpo Air Base, South Korea: (1) Wing Commander's Quarters, (2) Officers Quarters, (3) Eighth Army Photo Interpretation Center, (4) 67th Reconnaissance Technical Squadron Photo Interpretation Section, (5) 67th Reconnaissance Technical Squadron Photo Lab, (6) 67th Reconnaissance Technical Squadron Operation, (7) Kimche Bowl.

The 67th Reconnaissance Technical Squadron Photo Interpretation Section, Kimpo Air Base, South Korea, 1951.

PHOTO INTERPRETATION "C" SHIFT

The 67th Reconnaissance Technical Squadron Photo Interpretation Section "C" Shift, Kimpo Air Base, South Korea, 1951. *Front row, sitting, l to r:* Corporal V. W. Handley, Jr., and canine friend; Corporal M. Taylor. *Middle row, kneeling, l to r:* Staff Sergeant R. N. Brown; Corporal I. Colon; Technical Sergeant J. Pollack; Sergeants R. J. Oberley, D. K. Hitson; Staff Sergeant Duane Hall. *Back row, standing, l to r:* Captains T. E. Pippen, O. H. Milam. *Not pictured:* Captain J. G. Byers; Staff Sergeants D. M. Welcher, W. R. Grant; Sergeants R. R. Milota, H. W. Fuller, H. Wease, J. E. Gallaher, L. A. G. Solomon; Corporals F. A. Morgan, R. Nuzzetti, R. R. Ramos, L. C. Mahaffy; Privates First Class J. Salcido, A. R. Hayward, P. E. Papantonakis.

The 67th Reconnaissance Technical Squadron Photo Lab Section, Kimpo Air Base, South Korea, 1951.

The 67th Reconnaissance Technical Squadron Photo Lab "A" Shift, Kimpo Air Base, South Korea, 1951. *Front row, sitting, l to r:* Corporal M. Stidham*; Sergeant M. D. Jones; Corporals C. H. Veltsos, H. E. Trautner*, L. E. Erne; Technical Sergeant D. H. Blackman; Sergeant R. C. McCourty; Staff Sergeant J. F. Harper; Corporals T. J. Gits, M. J. Ruffins*. *Middle row, kneeling, l to r:* Master Sergeant M. C. Smith; Staff Sergeant W. D. Corley; Specialist First Class J. E. Gardner*; Corporal B. D. Raab; Staff Sergeants F. Fuduli, J. W. Mikeal; Privates First Class C. C. Graham, P. E. Nelson*; Sergeant W. E. Zimmerman; Staff Sergeant J. R. McMullan; Corporal T. S. Showlund*; Private G. H. Reding*. *Back row, standing, l to r:* Private First Class J. F. Kelahan*; Staff Sergeant L. A. D'Albertis; Sergeant A. M. Kaye; Staff Sergeant H. E. Karklin; Sergeant J. F. Mills; Staff Sergeant W. W. Teats; Technical Sergeant P. W. Somervold; Private R. F. Fee; Technical Sergeant G. P. Fletcher; First Lieutenant C. N. McCracken; Corporal J. R. Acker*; Staff Sergeant R. A. Neeb*; Privates J. W. Brown*, T. L. Berger*; Private First Class H. L. Williams*; Corporals H. R. Feflie*, J. C. Sellen*, L. E. Behn*; Private First Class M. S. Nemes*. *Not pictured:* Technical Sergeants F. N. Loper, W. H. Fairchild; Staff Sergeants J. L. Heap, H. E. Clark, P. Daskol, K. W. Brannon; Sergeants R. J. Lane, W. F. Schaffer, A. N. Fleshman*, R. D. LaCourse, P. J. Bolt; Corporals K. A. Erdman*, J. K. Bagley*; Privates First Class G. S. Joyce, W. M. Miyamoto*. (**Indicates 8199th Engineer Photography Reproduction and Distribution personnel.*)

67th Reconnaissance Technical Squadron Supply Section, Kimpo Air Base, South Korea, 1951. *Front row, kneeling, l to r:* Staff Sergeants A. L. Barfoot, R. W. Wentworth, H. S. Beltran; Corporals J. R. Bradford, J. M. Cameron. *Back row, standing, l to r:* Staff Sergeant M. L. Counts; Private First Class B. Rappaport; Technical Sergeant O. W. Steigerwald; Captain R. L. Burkett; Technical Sergeant C. W. Smith; Private First Class C. R. Seifman.

67th Reconnaissance Technical Squadron Maintenance Section, Kimpo Air Base, South Korea, 1951. *Front row, kneeling, l to r:* Sergeant R. J. Simmons; Staff Sergeant A. S. Frazier; Corporal G. W. Duncan; Privates First Class P. J. Elam, H. J. Dezotell, R. E. Canfield; Staff Sergeant P. T. Moore. *Back row, standing, l to r:* Private First Class L. L. Skold; Corporal R. Lumbert; Private First Class J. W. Grace; Corporal P. W. Loring; Sergeant C. R. Mullins; Staff Sergeant L. J. Beaulieu. *Not pictured:* Chief Warrant Officer R. E. Thompson; Master Sergeant H. E. Moss; Technical Sergeant R. E. Clark; Staff Sergeants W. W. McCluskey, A. N. Laxton; Sergeants R. Klatkiewicz, C. E. Nelson, J. L. Sebrell; Corporal J. L. Strohacker; Private First Class D. J. Day; Private R. C. Moss.

MAINTENANCE SECTION

67th Reconnaissance Technical Squadron Administration Section, Kimpo Air Base, South Korea, 1951. *Front row, sitting, l to r:* Staff Sergeants J. T. Stroman, C. L. Beaver; Technical Sergeant W. E. Bretz. *Back row, standing, l to r:* First Lieutenant S. A. Sosnow; Master Sergeant J. J. Balint; Private T. E. Lefort; Corporal W. H. Hiraga; Private First Class D.J. Watson. *Not pictured*: Captain T. M. Tarplay; Technical Sergeant C. M. Moody; Staff Sergeant K. P. Larsen; Sergeant E. Dabkowski; Staff Sergeant H. F. Gilg; Corporal J. P. Boyles.

Major Arthur A. Grumbine, Officer in Charge, Photo Interpretation.

Captain Thomas J. Evans, Officer in Charge, Operations.

Captain Wallace E. Brunson, Officer in Charge, Photo Lab.

Captain Robert L. Burkett, Officer in Charge, Supply.

Captain Eugene C. Cheatham, Jr., Officer in Charge, Intelligence.

Captain Thomas M. Tarplay, Adjutant.

Chief Warrant Officer Robert E. Thompson, Officer in Charge, Maintenance.

67th Reconnaissance Technical Squadron Staff Officers, Kimpo Air Base, South Korea, 1951.

2,177,000 Aerial Pix Produced In 5 Months

67TH TAC RECON WING, Korea.—Photographic laboratory technicians of the 67th Tactical Reconnaissance Wing, assisted by members of an Army photo team, produced 2,177,000 aerial photographs of North Korea during a five-month period in support of tactical air power in Korea.

Arriving in Korea during the month of September 1950, the advanced Fifth Air Force unit set up its processing laboratories in an abandoned Korean school house several miles from the battle lines. During the fight for the Pusan perimeter, the photo lab and intelligence sections supplied UN air and ground forces with photographic reproductions of Communist controlled strongholds and supply installations.

"Our men have turned out enough photographic prints in this war to make a nine-inch-wide highway from the docks at Pusan to the Yalu River and far into Manchuria," said Maj. Arthur A. Grumbine, officer-in-charge of the units' photo intelligence section.

T/Sgt. Howard Scott, shift chief, said: "Besides making prints, we have also processed over 40 miles of film, covering every known marshalling yard and supply dump north of the 38th Parallel."

"A lot of credit has to be given to the laboratory men who during the winter worked with coats on and their hands in near-freezing water and in the summer worked in the dark rooms with temperatures up to 120 degrees," said squadron commander, Lt. Col. Joseph Condon.

From *Air Force Times*, September 1, 1951.

Above: 67th Reconnaissance Technical Squadron Operations Section, Kimpo Air Base, South Korea, 1951.

Front row, sitting, l to r: Staff Sergeant N. V. Brown; Sergeant M. Techner*; Corporal J. R. Donham; Sergeants C. M. Rodgers, G. H. Simpson. *Middle row, kneeling, l to r:* Staff Sergeant R. D. Young; Sergeants C. H. Waldron*, L. W. Levy, J. P. Jones; Corporals V. H. Raaen*, J. O. G. Beaudoin. *Back row, standing, l to r:* Technical Sergeants F. R. Pfeifenberger, W. N. Meadows; First Lieutenant R. R. Long; Captain T. J. Evans; Staff Sergeants W. W. Polenz, H. R. Marsh. *Not pictured:* Master Sergeant C. Glew; Staff Sergeant F. G. Canedo; Privates First Class R. J. Pelletier, R. E. Lorenz. (*Indicates 8199th Engineer Photography Reproduction and Distribution personnel.)

of the Army Omar N. Bradley, accompanied by General Matthew B. Ridgway, Commander of the Allied Ground Forces. This gave the 67th another opportunity to brief the higher echelon on the air-fields and aircraft status in North Korea, the People's Republic of China, and the USSR from the very large small-scale Situation Map maintained on the wall between Lieutenant Ben Hardy and Staff Sergeant Duane Hall's work area. Praise from General Bradley was greatly appreciated, along with a comment of thanks from General Ridgway for the support provided the Eighth Army by the 67th RTS.

Another distinguished guest was Air Vice Marshal Knox from the Royal Australian Air Force, School of Air Warfare.

L to r: General of the Army Omar N. Bradley and General Matthew B. Ridgway (hidden by Bradley) are welcomed by Colonel Vincent Howard, 67th Tactical Reconnaissance Wing Commanding Officer (partially hidden by Lieutenant Colonel Schuyler S. Harris, Commanding Officer of the 67th Reconnaissance Technical Squadron).

Above: Visiting Secretary of the Air Force Thomas K. Finletter views airfield identification photographs shown by then Second Lieutenant Benson B. Hardy, 67th Reconnaissance Technical Squadron (67th Tactical Reconnaissance Wing, Fifth Air Force) Korea.   *USAF photo*

Left, l to r: General Omar N. Bradley; unknown; General Matthew B. Ridgway; and Lieutenant General James van Fleet arrive at Kimpo Air Base, Korea, summer 1951. The purpose of the visit was to inspect the 67th Tactical Reconnaissance Wing and its subordinate units, which included authors Ben Hardy and Duane Hall's 67th Reconnaissance Technical Squadron.

Being close to the front line had advantages as well as disadvantages. On the Situation Map, the 67th RTS maintained markings indicating bombing targets, along with the advance position and the airfield status; this made the map highly classified, particularly with the airfield status in Manchuria and Siberia annotated. Very often the unit had visits from front-line officers requesting aerial photo coverage of their area of concern from the negatives in the 67th film library. During these visits the officers would sometimes be able to update the current front-line status on the 67th's chart. This subsequently might enable members of the group to make return visits to an area, and that gave the 67th RTS a better understanding of how to support the men at the front.

One U.S. Army "Bird" Colonel (GI parlance for a full-grade Colonel, versus a Lieutenant Colonel, derived from the eagle insignia designating a Colonel) in particular became a regular visitor to the 67th Reconnaissance Technical Squadron Photo Intelligence Department. His usual procedure was to check

Visiting the 67th Reconnaissance Technical Squadron, Kimpo Air Base, summer 1951, are (*l to r*) General Matthew B. Ridgway, Supreme Commander, Allied Powers, and General of the Army Omar N. Bradley, Chairman of the Joint Chiefs of Staff. Colonel Vincent Howard is at the right. The photo was taken in the work area of Lieutenant Ben Hardy and Sergeant Duane Hall. Generals Ridgway and Bradley are viewing Composite Defense Report No. 5.

Lieutenant Ben Hardy reviews the airfield status in North Korea, Manchuria, and Siberia with Lieutenant General Frank F. Everest, Commander, Fifth Air Force.

Lieutenant General Frank F. Everest congratulates First Lieutenant Benson B. Hardy at Seoul, Korea, 1951, on his award of the Bronze Star for service as a photo interpreter.

on the unit's Top Secret Situation Map, which was updated daily. He was sometimes amazed at the accuracy portrayed, and he also helped the men with updating his sector. On one occasion he needled them on the "plush" conditions the "Air Force people" worked in, compared with what he and his troops had to contend. He also commented that he

did not understand why the Air Force "jet-jockeys" could not get down low enough to obtain photo recon that would be of value to him. And that was where the Colonel had almost overstepped his bounds. He had underestimated the courage and skill of the USAF RF-51 Mustang and RF-80 Shooting Star photo recon pilots. It was obvious he was having a bad day. After he tired of growling at the men, he looked at Sergeant Hall and announced, "Say, Flyboy, why don't you come up with me and I'll show you how this war is really fought!" Viewing his comment as a challenge, the Sergeant accepted the offer. He looked across at Lieutenant Hardy, who queried, "Are you going out for coffee, Sarge?" The reply came back: "Yes, Sir!" Donning his field jacket and picking up his carbine and helmet, Sergeant Hall told the Colonel, "Let's go!"

The jeep ride to the Colonel's command took about 45 minutes. About midway, the Colonel looked at Hall and asked, "Are you getting nervous, Flyboy?"

"No, Sir!" was the quick answer.

Actually Sergeant Hall felt quite privileged to be riding in the Colonel's Jeep.

At their arrival at the command post, Hall noted how quiet it was for a front-line area. That was soon to change. The Colonel had gathered some of his officers and briefed them on his operational orders. Hall was totally awe-stricken at what he could see: tanks, self-propelled guns, artillery pieces, and quad .50-caliber machine guns — four guns mounted on a half-track or trailer — lined up for probably a quarter of a mile. Then, at the Colonel's command, all hell broke loose. Hall and the Colonel were situated

Second Lieutenant Ben Hardy and RAF Flight Lieutenant Reginald L. A. Roberts check details on a reconnaissance photo at Kimpo Air Base. The USAF photo was published in the October 3, 1951, *Stars and Stripes*, noting the "close-knit teamwork of the United Nations" as the two men "go through their daily job of scanning aerial photographs of communist installations. The two work together on the Far East photo intelligence teams."

Sergeant R. E. Kennedy, photo interpreter with the 67th Reconnaissance Technical Squadron, at the squadron sign, Kimpo Air Base, 1951.   *Al Bertin photo*

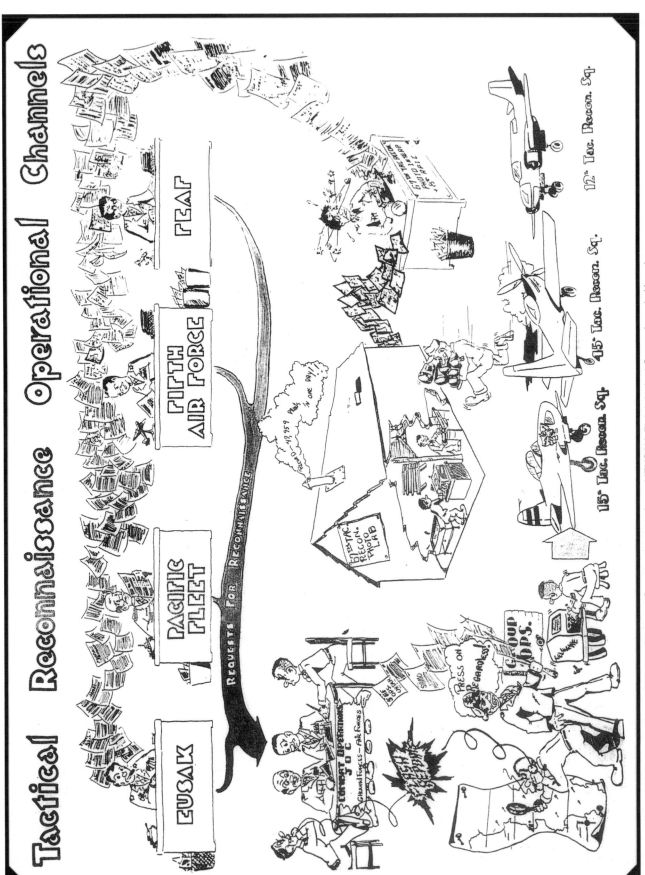

The work-flow chart from the 67th Reconnaissance Technical Squadron's 1951 *Annual* (EUSAK: Eighth United States Army in Korea).

at the foot of a long ridge, and when all of this eruption commenced, the earth shook. The explosions were so loud Hall wondered if he would ever be able to hear again. Those "ground-pounders" literally tore the side of the hill to pieces. After what seemed like an eternity — though it was about 30 minutes — the firing ceased. The silence was stunning. The Colonel came over to the jeep and asked if Hall wanted to stay for the next phase of the operation. The Sergeant told the Colonel he had seen what he had wanted to see, and that he had, so to speak, unknowingly written another chapter in the book of his life. The Colonel told his driver to "Take the Flyboy back to Kimpo . . . so we don't get him hurt!" Then he surprised Hall. He put out his hand and added, "Sergeant, you're all right!" Hall shook the Colonel's hand and replied, "Colonel, you're welcome at our place anytime."

33 ON 15TRS R6192A 5AF 22OCT51

An extremely low-altitude forward oblique photo in response to a U.S. Army Colonel's request for better coverage of his area. The mission was flown by the 15th Tactical Reconnaissance Squadron on October 22, 1951.

Hall had been gone about three hours, and when he arrived back at Kimpo, he returned to his work area with a big grin on his face. With an equally big grin, Lieutenant Hardy offered, "It must have been pretty good coffee!" The two men talked at some length about what the Colonel had said about needing low-level photo recon, and Lieutenant Hardy then spoke with Major Grumbine about it. The Major subsequently visited the 15th Tactical Reconnaissance Squadron to see what could be done.

The 15th TRS decided that if the Colonel wanted low-level reconnaissance, they were the people who could provide it. The assigned pilot, First Lieutenant Richard McNulty, stopped by the Photo Intelligence Department, and Lieutenant Hardy and Sergeant Hall briefed him on the area of concern. The next day, October 22, 1951, photos taken at almost ground level, from mission R6192A, were delivered for review and analysis. With these aerial photos the photo interpreters were able to detect bunkers and machine-gun emplacements that had not been picked up on the usual vertical photography. Bunkers with corrugated steel on top of them, and covered with dirt and tree limbs, were now readily visible.

The photos were taken at such a low altitude that it was impossible to accurately plot them on a map. When Lieutenant McNulty came over to see the results of his mission he was extremely proud of the photo coverage of the target area he had been assigned. Even though the photo interpreters couldn't accurately plot the photos on a map, all involved concurred that he had 100 percent excellent coverage of the mission.

A couple of days later, the unit's "favorite" Colonel came back for a visit. It should be noted at this point that the men of the Photo Intelligence Department did not view him as a bad sort. The eagles on his collar indicated he had experienced the full scene of World War II, and they had earned him the privilege of growling at anyone he wanted to that was not wearing a star.

After the Colonel had inspected the unit's Situation Map once again he acknowledged, "You guys are pretty damn accurate." When he asked if anything new was going on in the war, Lieutenant Hardy responded, "As a matter of fact, Colonel, there is." At that point the men spread out the photos of the low-level mission, which the Colonel had more or less "requested."

Following the discussion about that mission and other photo-intelligence information the squadron provided to all services concerned, the Colonel again praised the 67th's work. "You know, for Flyboys, you guys are pretty damn good at what you do." He thanked the men for their help. Regretfully, the 67th RTS never saw him again after he left with that set of photos. But another dimension had been added to photo interpretation via this special "backdoor" favor to the Colonel. In those early stages of the war, with the confusion of calls for reconnaissance missions and to enable the photo interpreters to bring order out of the chaos, all requests from front-line Commanders for such missions normally had to go through the Eighth Army planning group for approval, and then on to the Fifth Air Force Planning Group. But the 67th RTS had circumvented that to fill the Colonel's "unofficial" request.

Another interesting aspect of the photo intelligence experience was to be a "ride-along" passenger aboard an RB-26 Invader on a night reconnaissance mission. In order to stop the flow of supplies and personnel to the North Korean and Chinese armies, Far East Air Forces and the Fifth Air Force had once again instituted a high-priority interdiction program on the railroad bridges, railyards, and the highway bridges. This program was a real challenge to the photo interpreters trying to locate the camouflaged vehicles and the status of the bridges. As part of the program, the photo interpreters also had been keeping a surveillance check on some of the main arterial railroad bridges. Those that held their particular interest were at Sunchon. They had suspected, that during the night, standby replacement spans were being set up on one of the bridges that they had been reporting as "unserviceable," and thus the enemy was still moving its freight trains after dark. A request was therefore put in for night coverage of the bridges. After an approval had been received, the assigned pilot met with Lieutenant Hardy in the PI Department for a briefing on the site. Following the briefing, Captain Lyman Beck, in jest, asked Hardy to ride along with him. To his surprise, the Lieutenant accepted the

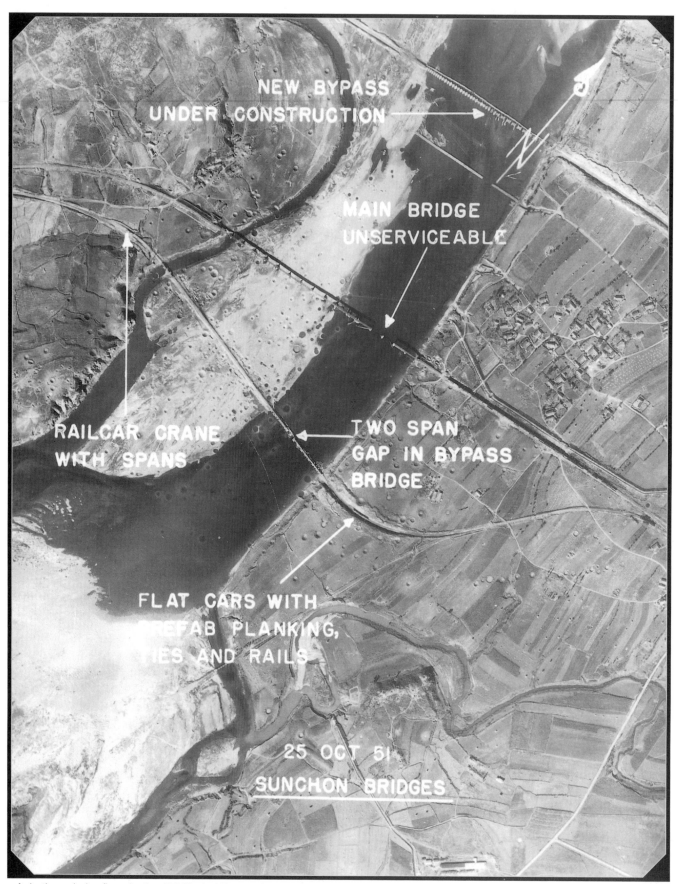

A daytime mission flown by the 15th Tactical Reconnaissance Squadron, October 25, 1951.

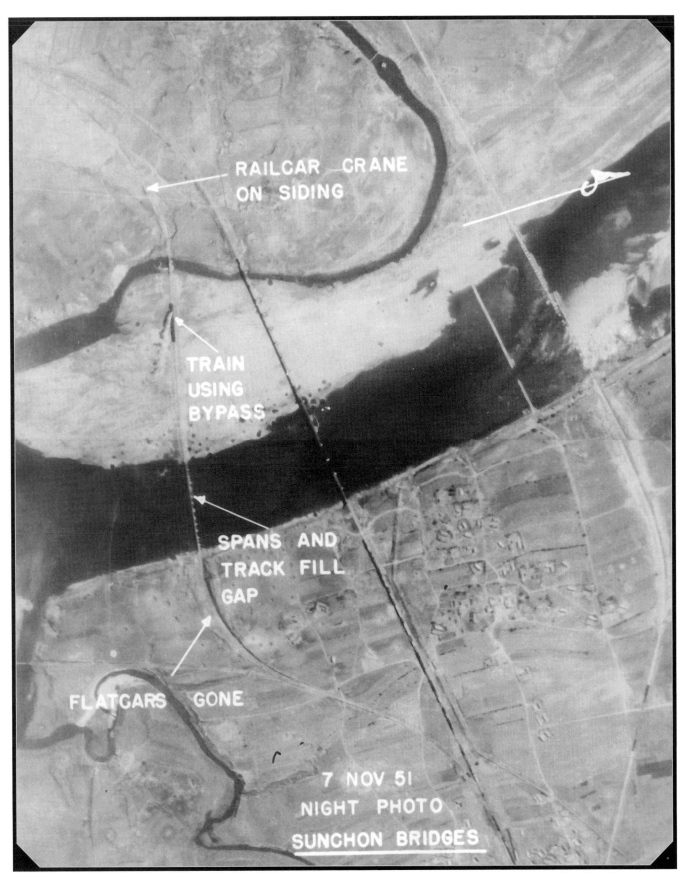

RAILCAR CRANE
ON SIDING

TRAIN
USING
BYPASS

SPANS AND
TRACK FILL
GAP

FLATCARS GONE

7 NOV 51
NIGHT PHOTO
SUNCHON BRIDGES

Proof of the suspected enemy night movement of railcars from the mission flown by the 12th Tactical Reconnaissance Squadron, November 7, 1951.

challenge. Hardy recalls the exciting part of the flight was when some "gooks" (as the South Koreans called the men of the North Korean People's Army) fired a few rounds of antiaircraft in their direction, though not close enough to cause great concern. When Lieutenant Hardy and Captain Beck arrived over the bridge site, Beck released some flash bombs, and his cameras clicked off several shots. With the mission objective completed, they headed back to Kimpo Air Base.

Above and below: Rather ineffective North Korean attempts at camouflaging rolling locomotives, photographed in July and August of 1950.

The film was delivered to the photo lab for the processing and printing, while Captain Beck and Lieutenant Hardy went over to the PI Department to wait. After what seemed like a long and anxious time, the prints were finally delivered to them. They already knew what was going to show up, as they had spotted a train that had just crossed over the bridge in question during the reconnaissance run. Yet Lieutenant Hardy says it was an exhilarating experience to write a Mission Review Intelligence Report on which he had flown as a "ride-along."

The photos proved out what had been suspected, and soon a strike mission was sent on its way to take care of the matter. The next photo coverage, they hoped, would show that the target had been obliterated.

Fortunately, those who had experienced one of these "ride-along" excursions all returned none the worse (although not so for some unlucky crews), and they had a better understanding of the hazards faced by an unarmed reconnaissance pilot and crew. These unauthorized and unorthodox "on the job" training sessions at times produced valuable results. Ride-along missions in RB-26s and short visits to the front-line positions produced a better insight into the information the photo interpreters could provide for both air-combat and ground-force operations.

It was rewarding to know that the intelligence reports and photographs from the Photo Interpretation Department were in such a demand for their details and accuracy. Many interesting items were found when studying the aerial photographs, from which the PIs were writing their reports. In late May 1951, during his debriefing, a fighter pilot reported seeing the Russian-built MiG-15 fighter he shot down "clobber in" without burning. The 67th TRW was immediately notified and was requested to fly a reconnaissance mission over that area to locate the enemy plane. At the time, the Air Force had no details of the MiG-15, though in December of 1950 it had become evident, with air-to-air combat in the 27th Fighter-Escort Wing, that the American F-84D Thunderjets were inferior to the Russian craft. The 67th's mission was flown and photos of the downed MiG-15's location were then plotted on a large-scale map. On June 1 a unit parachuted into the site to retrieve parts from the aircraft, which would help the pilots to understand what they were up against.

Much interesting information was found when studying the aerial photographs from which the PIs were writing their reports. The North Koreans went to great lengths trying to hide their Polikarpov

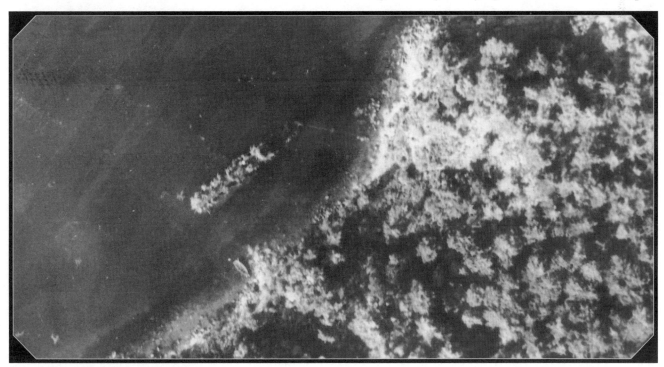

A camouflaged North Korean ship offshore.

PO-2 aircraft, an antique biplane used to harass the Americans at night, flying at low altitudes below the radar coverage. The PO-2s — what historian Robert Futrell called "night hecklers" — flew over U.S. installations, dropping hand grenades and/or mortar shells. The North Koreans hid the PO-2s in small villages and used the roads as airstrips. They also built a dummy airfield, complete with dummy aircraft revetments and dummy aircraft included.

Another ruse that the photo interpreters detected was the use of dummy bomb craters, first noted by Lieutenant Hardy. Following an attack on their major airfields, the Communists would fill in and repair the bomb craters, after which they would build circular mounds of dirt or sand bags around the repairs, giving the appearance that the runway was still unserviceable. During the intense study of what the photo interpreters were reporting as unusable runways, Lieutenant Hardy noticed that the craters were casting shadows where there should not have been any. Robert F. Futrell, in *The United States Air Force in Korea 1950-1953* (1983, rev. ed., pp. 417-418), later wrote concerning the newly constructed airfields in the MiG Alley area:

The increased intensity of the Far East Air Forces interdiction program against the Communist railroad and truck resupply efforts to their front-line armies forced the enemy to move its trains and trucks at night. During the day, trains would be hidden in tunnels and trucks camouflaged and parked alongside roads, as seen above with the five marked vehicles. The photo was taken on April 26, 1951, mission R3578C, flown by the 15th Tactical Reconnaissance Squadron.

After a USAF fighter pilot had reported he had shot down a MiG-15, and saw where it clobbered in without burning, a reconnaissance mission was flown to locate the downed aircraft. This photo, from that mission, shows the MiG-15 in the center. The low altitude of the reconnaissance aircraft produced a somewhat fuzzy photograph.

Above: A Russian MiG-15 fighter being pursued by a North American F-86 Sabre Jet.

Above and left: F-86 Sabre Jet aircraft at Kimpo Air Base, Korea, 1951.

By guise and by guile the Communists attempted to counter the night-flying bomber attacks. Evidently hoping to confuse the B-29 crews, the Reds piled circular rings of dirt on the runways at their MIG [*sic*] Alley airfields to simulate bomb craters. A sharp-eyed FEAF photo interpreter almost immediately noted that the dummy bomb craters were not the right size, and low-level reconnaissance verified that the craters were piles of loose earth, banked up on unharmed sections of the runways.

The photo interpreters did not believe that the dummy bomb craters were meant to deceive the night-flying, high-altitude B-29s, but rather to fool the PIs or the daylight-flying strafing attacks into believing the installation was unserviceable. At this time of late 1951, the B-29s were flying their bombing missions at night because of the MiG-15s attacking out of Manchuria, and at their high altitudes, even in the daylight, the B-29s would never be able to discern any dummy bomb craters.

The reporting of the dummy craters caused quite a controversy in FEAF intelligence circles. An

Two F-86 Sabre Jets taking off from Kimpo Air Base, Korea, 1951.

Known as "Radar Mountain" in official channels, this summit was located west of Kimpo Air Base. The powerful radar, and at that time the latest available, was situated on this mountain top. The capabilities of this installation, along with the one on Cho-do Island, southwest of Chinnampo, North Korea, were primarily responsible (along with superb pilot skill) for the outstanding kill ratio of the F-86 Sabre Jets compared with the Russian MiG-15s. The GIs named this mountain "The Witch's Tit."   *Lloyd Wooley photo*

officer in another unit wrote a report that was printed in the FEAF Intelligence Summary alleging the craters to be a "figment of Lieutenant Hardy's imagination." Lieutenant Colonel Joe Condon then instructed Lieutenant Hardy to do an in-depth detailed report on the dummy bomb crater controversy. Low-altitude forward oblique photos provided by the U.S. Navy, and the 67th RTS's own file photos, helped to establish the fact that the

North Koreans were indeed using this subterfuge. Then later, the re-interrogation of a defector from the army of the People's Republic of China further substantiated Hardy's discovery of the dummy bomb craters.

Yet the attempted camouflage was detected much earlier on other airfields. The North Korean People's Army and Chinese manpower could remove any camouflage during the night and bring in

Camouflaged enemy aircraft found at Pyonggang Airfield, North Korea.

serviceable aircraft, then hit the Allied Forces with a surprise attack. However, with the detection of this ruse, Bomber Command scheduled a program to continually "post-hole" those airfields, and thus the Communists were never able to put them into sustained operation.

Another fascinating discovery by a squadron intelligence officer/photo interpreter, Captain Gene Cheatham — one of the intrepid Tuskegee Institute airmen, who was always slipping out to the airfield to persuade one of the reconnaissance squadrons into letting him fly recce — was the discovery of a

set of large-scale forward-oblique photos that showed the Chinese using camels to haul supplies and equipment. Occasionally, the photo interpreters were requested to select some of these strike and post-strike photos to be used in producing psychological-warfare propaganda leaflets.

But perhaps the most memorable incident of enemy subterfuge involved an unmarked POW encampment close to the Yalu River. The facility appeared to have been a former school with a large empty field adjacent to the building. In this open area, the POWs were forming together to spell out

This photo shows a futile attempt by the North Koreans and the Chinese military to deceive the photo interpreters and attack aircraft by building a dummy airfield. Included are dummy aircraft, dummy aircraft revetments, a dummy operations building, a dummy vehicle, and a dummy. It didn't fool anyone. Adding his own wry humor, one of the USAF line crewmen had written with chalk on a bomb, *"Hey dummy, this ain't no dummy."* A USAF pilot willingly delivered it — mission R7346A, flown December 24, 1951, by the 15th Tactical Reconnaissance Squadron.

HEADQUARTERS
FIFTH AIR FORCE
Office of the Air Facilities Branch
APO 970

6 November 1951

MEMORANDUM FOR: General Everest

SUBJECT: Prisoner of War Special Interrogation Report

1. Re-interrogation of Chinese PW Hu He Hsing, was conducted by Lt Col Robert C. Smith, Director of Photo Intelligence and Captain Leonard F. Nielson, Chief, Air Facilities Branch, this headquarters, on 4 November 1951. The primary object of the re-interrogation was that of obtaining additional information relative to the construction of Sunan #2 Airfield and simulated craters noted thereon. PW was above average intelligence; of good memory, and proffered information in an unhesitating manner.

2. <u>Personal details of PW</u>

PW Name:	Hu He Hsing
PW Number:	CD 13,176
Age:	22
Education:	3 yrs
Rank:	Pvt
Duty:	Ammo bearer
Place of Capture:	CT 255304 (Deserter)
Date:	30 Oct 51
Unit:	CCF, 47th Army, 139th Div, 417th Regt, 2nd Bn, Hws Co, 3rd Plat, 5th Squad.

3. <u>Construction of Sunan #2 Airfield:</u>

During the months of April and May 1951, PW and his Company performed construction work at Sunan #2 Airfield. PW stated that all basic construction work, prior to the first B-29 strike, was performed by recruited civilians from the town of Sunan and its surrounding environs, however, after the first B-29 strike, civilians concerned refused to participate further in its construction. At this juncture, PW states an army division, exact strength unknown, was moved in to complete the airfield. PW heard that the surveying of the airstrip and other engineering problems were accomplished by Russian technicians. PW states that the construction took place during the hours of darkness (8-10 hrs) and that the division labor was divided into groups of personnel, each assigned to and responsible for, the completion of a certain area of runway.

Above and opposite: The Prisoner of War Special Interrogation Report that confirmed the photo interpreters' observation of the enemy's use of dummy bomb craters in North Korean airfields that in reality had been repaired.

No lighting facilities were used. The only type of heavy equipment noted by the PW consisted of a roller. According to the PW, the runway composition consisted of a base course of rock, of varying sizes, approximately eighteen (18) inches in depth and capped with concrete eight (8) **to** ten (10) inches in thickness, and without the benefit of any sort of reinforcement materials. Wood expansion joints were used and were left intact upon completion of each section of runway. PW states that after the pouring of concrete had been completed, a thin layer of sand was evenly applied as camouflage. Note: It is more likely that the application of sand, at this stage of construction, was intended to promote the curing of the concrete surface. PW recalls that the cement used at Sunan #2 Airfield arrived by train in <u>unlabled</u>, double thickness paper bags, and were transported by truck from the rail road station to the airfield. The aggregate, which had been previously stockpiled along the runway proper was then quickly and efficiently mixed with the cement and poured. No delays were experienced at any time, due to shortages of cement.

4. <u>Simulated Bomb Craters</u>:

Simulated craters, according to the PW, were constructed almost immediately after completion of the runway or any portion thereof, and consisted of rings of sand and gravel approximately three (3) to four (4) feet in height. Soot and ashes were applied to the tops of these rings, occasionally, to further accentuate the simulated craters.

Further interrogation of subject PW revealed that after a strike by B-29 type aircraft, the area affected would be roped off and all construction activities would be suspended for three (3) days. No attempts were made, so far as he knew, to ~~either~~ de-fuse or remove the bombs. PW further stated that in some cases workers caused self inflicted wounds to fingers, hands and legs in an effort to be evacuated and thereby escape further duty of this nature. "Flying shrapnel", was the excuse given for the wounds inflicted.

LEONARD F. NIELSEN
Captain, USAF
Chief, Air Facilities Branch

An enclosure photo of Ongjong-ni Airfield, North Korea, May 2, 1951, from the Special Report on Dummy Bomb Craters. The numbers on this photo correspond with these on the photo on the facing page (Ongjong-ni Airfield, Photo 1). The mission was flown by the 15th Tactical Reconnaissance Squadron.

ONGJONG-NI A/F
XD 939078
5 AF TGT. NO. A-771
DUMMY BOMB CRATERS
67 RTS 2 MAY 1951
PHOTO 4 BBH

ONGJONG-NI A/F
XD 939078
5AF TGT NO. A 771
DUMMY BOMB CRATERS
67RTS 6 MAY 51
PHOTO 1   BBH

This oblique photo of the dummy bomb craters on Ongjong-ni Airfield was provided by a U.S. Navy flight of May 6, 1951. (Numbers correspond to Photo 4, May 2, 1951, on preceding page.)

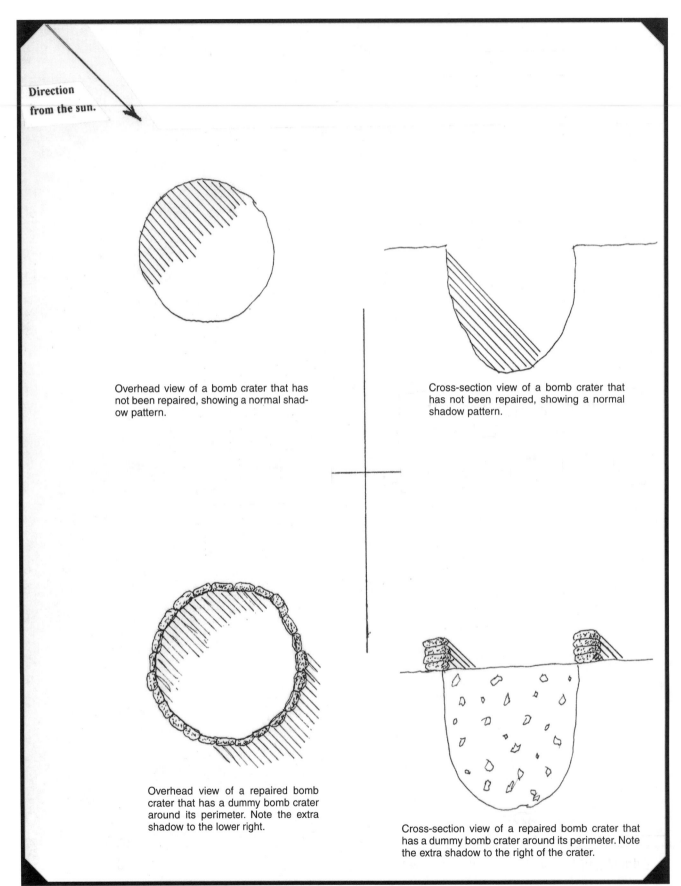

Direction from the sun.

Overhead view of a bomb crater that has not been repaired, showing a normal shadow pattern.

Cross-section view of a bomb crater that has not been repaired, showing a normal shadow pattern.

Overhead view of a repaired bomb crater that has a dummy bomb crater around its perimeter. Note the extra shadow to the lower right.

Cross-section view of a repaired bomb crater that has a dummy bomb crater around its perimeter. Note the extra shadow to the right of the crater.

Diagrams showing clues to detecting dummy bomb craters.

By Gene Leicht

their status — POW — by standing in a pattern. They had already created the P and the O and were still in the process of forming the W when the reconnaissance aircraft flew over. Evidently the POWs had realized that it was an Allied aircraft over their compound. (Unfortunately, efforts to retrieve photographs of this incident from the National Archives have proven futile.) The POW compound had no visible markings on the building or the ground indicating the camp's status, as was required by the Geneva Convention. But, of course, North Korea has never been a signatory to that document. However, the PIs did find cases where the Communists were marking some of their military installations with a large cross, much like the Red Cross symbol, hoping to keep them from being attacked. They figured correctly that the Allies would abide by the Geneva Convention and consider such institutions medical facilities.

Surveillance of enemy airfields along the east side of the Yalu River, known as MiG Alley, in the fall of 1951 was severely curtailed because of the vulnerability of the RF-80 reconnaissance aircraft to the faster and more maneuverable MiG-15s stationed at Antung Airfield, west of the Yalu River in Manchuria.

On November 9, 1951, the 15th Tactical Reconnaissance Squadron flew mission R6585B over three of the airfields in the MiG Alley area. This was the 15th's first coverage since the previous mission flown October 29. At that time, Uiju Airfield was still under construction, as noted in the eight-page Special Airfield Report No. 23 of December 29, 1951 (the first page of the report is shown herein on page 71).

At the conclusion of photo recon mission R6585B, the pilot went to the Wing Operations Center to be debriefed. He reported that he had observed quite a few aircraft on Uiju and Sinuiju Airfields, but none on Sinuiju Northeast Airfield. Wing Operations Center immediately notified the 67th Reconnaissance Technical Squadron Operations Department to give the roll of film from that mission the highest priority in processing. The 67th Squadron Operations directed the photo lab to process the contact prints on Type 9 waterproof

paper and not take the time required to dry them, but to just run a squeegee over them and deliver them to the Photo Interpretation Department. Major Art Grumbine, Officer in Charge of the 67th RTS Photo Interpretation Department, was also alerted concerning this highest priority mission. In turn, he alerted Lieutenant Hardy, OIC of the Special Airfield Section.

The photo interpreters were aware of the airfields they would be looking at, and thus in preparation they had pulled the fields' target folders. Several high-ranking officers other than the 67th Wing Commander and Squadron Commander arrived while we were waiting for the mission prints. We did not know many of them, but assumed that Major Grumbine had checked to see if they were authorized to be in our area.

Finally, the prints were brought from the photo lab in a soggy bundle. Lieutenant Hardy took the photos of Uiju Airfield, the first ones on the stack, and the remainder were handed to Sergeant Hall with instructions to work them over. To interpret wet photos was a completely new experience. It took a little juggling to arrange them to produce clear stereo vision. A quick scan on and around Sinuiju Northeast revealed no aircraft, so those photos were set aside.

Hardy and Hall began making notes on what they could see. Colonel Edwin S. Chickering, 67th Tactical Reconnaissance Wing Commander, was standing between the desks, instructing both men to give verbal reports and write the regular evaluation later. Lieutenant Hardy reported 14 MiG-15 aircraft sitting on the alert hardstand, 10 MiGs in aircraft revetments, and 2 MiGs moving down the runway, for a total of 26 MiG-15 aircraft on Uiju Airfield. Colonel Chickering then asked if Sergeant Hall had anything to add. "Yes, Sir," he replied. "On Sinuiju Northeast there are no aircraft at this time. On Sinuiju Airfield five YAK-type [*Russian Yakolev*] aircraft are on or adjacent to the hardstand, one YAK-type aircraft is sitting on the edge of the landing area, and four YAK-type aircraft are positioned in aircraft revetments, for a total of ten YAK aircraft at Sinuiju Airfield."

At that point, Colonel Chickering commented, "Gentlemen, the information you have provided will reverberate immediately from here to the White House, the Pentagon, and in fact to the

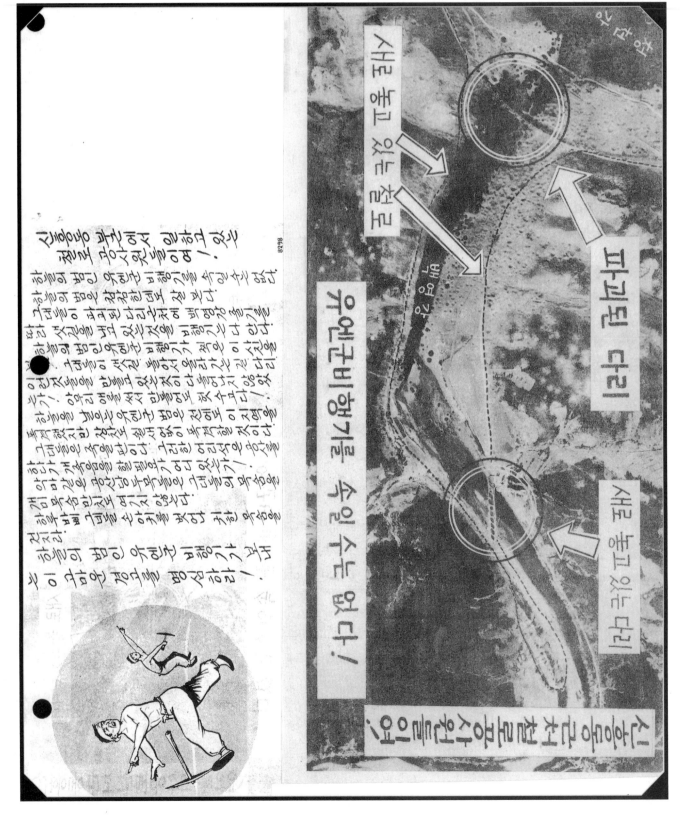

PSYCHOLOGICAL WARFARE DIVISION, G3
Headquarters, EUSAK
APO 301

7 June 1952

LEAFLET　　　　　:　Warning to RR By-Pass Workers

LANGUAGE　　　　:　Korean

DESIGNATION　　:　Serial Nr. 8298

TARGET　　　　　:　NK RR Workers in Sinhung-dong area

REMARKS　　　　:　Requested by Fifth Air Force. Aerial photography has re-
　　　　　　　　　　vealed that NK RR construction workers are building a long
　　　　　　　　　　by-pass around a destroyed bridge crossing the Paengnyong
　　　　　　　　　　River. Leaflet will tell them that this effort at deception
　　　　　　　　　　has failed and that the by-pass will be bombed.

ART WORK　　　　:　Vertical aerial photo showing by-pass construction.

(Obverse:　Photo with captions)

"RAILWAY BUILDERS IN THE SINHUNG-DONG AREA!"

"YOU CANNOT DECIEVE THE FLYING TIGERS OF THE UN!"

(Reverse:　Text)

"RAILWAY BUILDERS IN THE SINHUNG-DONG AREA!"

You cannot decieve the Flying Tigers of the UN!
They have eyes which can pierce the darkness!

They see that you are building a long by-pass
around the ruined bridge over the Paengnyong River!

Look at the photograph taken by Our Flying Tigers. Does it
not show where you are building the by-pass, switch-back, and new bridge?

Your work is useless! The UN Flying Tigers have
bombed this area before and will do so again and again!

Your lives are in danger! Is it not a pity to die a dog's
death for such useless work? Your evil Communist leaders cannot defend
you, so you must flee and preserve your precious lives.

HEED THIS FRIENDLY WARNING FROM THE UN FLYING TIGERS!

Translation of the propaganda leaflet opposite.

One of the reported compounds at Pyongyang, North Korea (1950), where Ameican POWs were being held. Note there are no markings on the buildings to identify the area as a POW encampment, as specified by the Geneva Convention.

67TH RECONNAISSANCE TECHNICAL SQUADRON
67TH TACTICAL RECONNAISSANCE GROUP
APO 970

29 December 1951

SPECIAL AIRFIELD REPORT NO 23.
UIJU AIRFIELD
40°09'N/124°29'E
XE 281453
Fifth Air Force Target No AXE2845

INTELLIGENCE SOURCES:

Photography:

Vertical and Tri-Met

Print Nos	Flying Organization	Mission No	Date	Scale
VV 28-32	8th Tac Recon Sq	R 1518-B	5 Nov 50	1:10000
VV 23-25	15th Tac Recon Sq	R 2955-B	19 Mar 51	1:10500
VV 15-16	15th Tac Recon Sq	R 3664-B	2 May 51	1:10000
VV 11-13	15th Tac Recon Sq	R 5041-B	18 Aug 51	1:10000
VV 8	15th Tac Recon Sq	R 5237-B	5 Sep 51	1:7500
RT 68-69	15th Tac Recon Sq	R 6116-B	17 Oct 51	Oblique
VV 17-22	15th Tac Recon Sq	R 6382-B	29 Oct 51	1:10000
VV 1-6	15th Tac Recon Sq	R 6585-B	9 Nov 51	1:8500
VV 21-27	15th Tac Recon Sq	R 6812-B	23 Nov 51	1:10000
VV 2-7	15th Tac Recon Sq	R 6983-B	1 Dec 51	1:10000
VV 4-9	15th Tac Recon Sq	R 7028-B	5 Dec 51	1:10000
VV 2-8	15th Tac Recon Sq	R 7120-B	11 Dec 51	1:10000
VV 12-18	15th Tac Recon Sq	R 7371-B	27 Dec 51	1:10000
VV 46-50	15th Tac Recon Sq	R 7371-B	16 Nov 51	1:10000

MAPS:

AMS SHEET 6134 III, Series L-751, Scale 1:50,000.

USAF Aeronautical Approach Chart 290 D IV, Scale 1:250,000.

INCLOSURES: (2)

1. Annotated Vertical Print
2. Annotated Map, Scale 1:50,000

SECURITY INFORMATION
Restricted

Page 1 of Special Airfield Report No. 23, an eight-page summary of missions describing the construction process for Uiju Airfield. This report erroneously lists the final mission date as November 16, 1951. The correct date cannot be verified.

United Nations. Due to the unusual circumstances, I sincerely appreciate the information you have provided. It is obvious you understood that time is of essence and you respected that."

Hardy and Hall's verbal report required approximately 15 minutes from the moment they had received the photos. All present that day were well aware that this was the Communists' first attempt to

station aircraft en masse on North Korean airfields this late into the war, as any aircraft they had previously tried this with had been destroyed at the start of the war.

Immediately, the Wing and Squadron Commanders departed, and when all were gone, the two men leaned back in their chairs, with a sigh of relief, and mentally digested the activity they had participated

This photo shows many of the 21 MiG-15 aircraft on the newly constructed Uiju Airfield in North Korea. The photo was part of those taken on November 9, 1951, by the 15th Tactical Reconnaissance Squadron, mission R6585B. Following the developing of the mission film, the 67th Reconnaissance Technical Squadron Photo Interpretation Department sent a "flash report" of the film findings to Fifth Air Force Headquarters, creating a mild panic.

National Reconnaissance Office photo

in. Soon Sergeant Hall looked over at Lieutenant Hardy and observed, "Lieutenant, I could use a beer about now to pull my nerves down off the wall." Lieutenant Hardy looked at him wide-eyed and jokingly responded, ". . . Lieutenants do not deliver beer to Sergeants!"

Sergeant Hall then replied, "I know that . . . but if they did, I would drink it!" A few additional inane comments broke the watch-spring tension.

The photos of the Uiju Airfield subsequently appeared as a double-page leading article feature in the December 10, 1951, issue of *Life* magazine. From November 9 on, frequent surveillance and attack missions were maintained against Uiju and Sinuiju Airfields, as well as other fields in the MiG

Sinuiju Airfield, North Korea. Ten Russian YAK (Yakolev)-type aircraft are visible in this photo taken November 9, 1951, by the 15th Reconnaissance Technical Squadron, mission R6585B.

National Reconnaissance Office photo

The Sinuiju-Antung bridges under attack, November 15, 1951, by the 98th Bombardment Wing. The Yalu River is in the foreground, and North Korea is in the upper portion of the photo.

Alley area — Kanggye #2, Namsi, Taechon, and Saamcham. (Interestingly, before the truce on July 27, 1953, no aircraft were seen on these airfields.) Constant attack and surveillance missions had to be maintained against the airfields in the MiG Alley area, as the Communists diligently continued to repair them. The results of these numerous attack missions were kept track of on the frequent surveillance reconnaissance missions. Only two MiG-15s were seen on Uiju Airfield on December 11, 1951, and on December 27 no serviceable MiG-15s were visible.

In an earlier issue of *Life* magazine, July 9, 1951, a color spread had featured oblique photos of two thoroughly destroyed installations in North Korea, the Wonsan Railroad Yard and the Wonsan Oil Refinery. The two photos were from the new type camera that was then being developed, the K-17 Sonne, which could take a continuous photograph, instead of a single 8x10-inch or 9x18-inch photo. A 390-foot-long photo could be accomplished by the pilot with the flick of a switch. However, processing of color film in the field at that time was a delaying factor in the camera's use.

On one occasion, an RB-26 mission's set of prints appeared on Staff Sergeant Hall's desk. As usual, with the pilot's plot in hand, he went to the library to get the aeronautical chart that covered the area indicated. Purportedly, the plot showed that the pilot had flown over Wonsan Airfield (K-25) on the east coast of North Korea. However, in checking the plot, and looking at the photos, Staff Sergeant Hall immediately recognized that the photos were of the "off-limits" Chinese airfield at Antung, Manchuria, approximately 175 miles northwest of Wonsan, across the North Korean border, at that country's far northwestern corner. He recognized the field from the previous oblique photographs of that area.

The "mis-identification" caused a big flap, as the pilot had flown the mission at night in an RB-26 from the 12th Tactical Reconnaissance Squadron using flash bombs to illuminate the target. It was a mission of vertical photography, which indicated the pilot had been flying over China to the west of North Korea. Though the pilot vehemently denied it, Hardy and Hall felt then that he had been directed to fly that mission; after all, the Chinese were already involved in the war. And 175 miles off

course was hard to believe! Knowing today about the many penetrating overflights during the Cold War period more than ever confirms their suspicions.

A similar RB-26 mission was accomplished over Vladivostok, USSR, where the pilot's purported target was conveniently cloud covered. The professed target was Hoeryong Airfield (K-35) in the extreme far reaches of northeastern Korea.

Sergeant Duane Hall was a "ride-along" on that mission. On May 2, 1951, he prepared and issued Special Airfield Report No. 18, Hoeryong Airfield, which included the facilities, conditions, and detectable activities of the entire field. His recollection of the surreptitious flight — "the mission that never happened" — follows.

The Mission That Never Happened

Hoeryong Airfield is located in the extreme northeast corner of North Korea, approximately 125 air miles from Vladivostok, Russia. North Korea, China, and, in fact, Russia had a strong desire to station aircraft within North Korea. The United States and the United Nations had a stronger desire to stop this from happening.

The mission of the Airfield Section of the 67th Reconnaissance Technical Squadron Photo Interpretation Department was to monitor all activities of all airfields in North Korea. Lieutenant Ben Hardy was the 67th RTS Officer in Charge, and I served as his non-commissioned officer. Lieutenant Hardy was given authority to request photo coverage of any field that he felt needed new photo coverage.

The date is uncertain, but in the afternoon around October 1, 1951, Lieutenant Hardy came into our work area and told me that he had requested aerial coverage of Hoeryong Airfield. He asked if I would like to make the mission as a "ride-along." Knowing that the Lieutenant had flown some of those same type missions, I felt this would be a great new experience. The mission would be at night, in an RB-26 Invader from the 12th Tactical Reconnaissance Squadron.

At dusk, we went to the flight line. Lieutenant Hardy introduced me to the pilot,

(handwritten) Restricted

67TH RECONNAISSANCE TECHNICAL SQUADRON
67TH TACTICAL RECONNAISSANCE GROUP
APO 970

28 October 1951

<u>SPECIAL AIR FIELD REPORT NO 21</u>

NAMSI AIR FIELD
39°57'N/125°13'E
XE 885215
Fifth Air Force Target No. AXE 8821

<u>Intelligence Sources:</u>

Photography:

Verticals.

Print Nos	Flying Organization	Mission No	Date	Scale
VV 53	15 Tac Rcn Sqdn	R-6034 C	14 Oct 51	1:22,600
VV 21-25	15 Tac Rcn Sqdn	R-6076 B	16 Oct 51	1:5000
VV 11-15	15 Tac Rcn Sqdn	R-6152 B	18 Oct 51	1:7,300

Maps:

AMS Sheet 6233 IV, Series L-751, Scale 1:50,000
USAF Aeronautical Approach Chart 380 A II, Scale 1:250,000

Inclosures:

1. Annotated Vertical Photo.
2. Annotated Map, Scale 1:50,000

-1-

(handwritten) Restricted

The Namsi Special Airfield Report.

SUMMARY:

The construction of NAMSI Airfield was first noted on photography flown 14 Oct 51. At that time, the complete outline of the R/W, 7100' X 250', the taxiway and the dispersal areas on the Southwest side of the R/W were plainly laid out and construction was well under way. The concrete had already been poured on the R/W and the base coarse had been packed on the taxiways. Track activity between the A/F and the creek bottom indicated that the base coarse for the R/W and taxiways had been procured from the creek bottoms. Vehicle revetments in the Southwest vehicle park area were already complete. In addition to the vehicle revetments there were twelve (12) open pits, each measuring 38' X 15'.

Photography flown 16 Oct 51 revealed a very rapid pace in the construction of the A/F. Almost all of the construction materials had been used or removed from both sides of the R/W. Construction of fifty-six (56) "U" type revetments was under way. Approx 50% of the surface of the taxiways were paved with concrete. There were fifty-one (51) large camouflaged objects, possibly vehicles, noted at the NW end of the R/W. These were camouflaged to appear as hay stacks. Numerous other vehicles were noted moving about on the A/F. Six (6) additional vehicle revetments had been constructed in the parking area Southwest of the A/F.

On cover flown 18 Oct 51, the workers had begun to pile the sand which had been laid over the R/W as a moisture retainer. This indicated that the concrete was cured and ready for use when the sand was removed. There were sixty-two (62) "U" type A/C revetments all but seven (7) of which were complete. Alert stands were being started at both ends of the R/W. Heavy equipment was being used to level the area where alert stands were being constructed. The concentration of personnel in the A/C revetment area indicated that attention was being focused on completion of the A/C revetments. The fifty-one (51) U/I camouflaged objects noted on 16 Oct 51 have all been removed.

DESCRIPTION:

1. Reference Point: R/P is the midpoint of the present R/W.

2. Location: The A/F lies in a small valley 3½ miles NNW of the town of NAMSI. The KUSONG Reservoir lies 7000' NW of the R/P. The town of KUSONG lies approx five (5) miles NNE of the A/F.

3. Altitude: Approx five hundred (500) feet above mean sea level.

4. Landmarks: There is a river fork five (5) miles due East of the A/F. The CHONGJU - KUSONG - SAKCHU rail line passes 1½ miles East of the A/F. The KUSONG Reservoir lies 1½ miles NW of the R/P. A smaller reservoir (unnamed) lies 4000' East of KUSONG Reservoir.

-2-

Restricted

Restricted

67 RTS, Spec Air Field Rpt #21, dtd 28 Oct 51, cont'd.

5. Flying Obstructions: There is a mountain peak, 1000' above sea level,
    approx seven (7) miles SSW of the A/F. Another range of mountains also
    1000' above sea level, lies six (6) miles West of the A/F. This range
    extends in a North - South direction. There are low foot hills on both
    sides of the A/F, ranging from 600' to 800' above sea level. There are
    no high tension power lines in the immediate vicinity of the A/F.

6. Runways: A single NW - SE runway, 7100' X 250' of concrete construct-
    ion has been poured. Photos of 18 Oct 51 show that the material used as
    a cover for curing the concrete being removed.

7. Hangers and Work Shop: There are no conventional type hangers or work
    shops visible.

8. Dispersal Area and Facilities: A taxiway approx 6000' long, parrellel
    with the R/W has thirty-eight (38) A/C revetments under construction.
    In addition there are two (2) taxiways extending to the South from each
    end of the R/W. Of these two (2), the NW taxiway is approx 4300' long
    with eleven (11) A/C revetments and the SE taxiway is approx 3700' long
    with thirteen (13) A/C revetments. There are six (6) vehicle revetments
    in a creek bank North of the A/F. A vehicle parking and supply storage
    area with at least twenty-nine (29) M/V revetments and twelve (12) open
    pits measuring 38' X 15' is located 5000' SW of the R/P. One (1) ammo
    revetment is under construction in close proximity of SW taxiway. Two (2)
    alert stand parking areas are under construction at each end of the R/W.

9. There is a POL storage area in the dispersal area 5400' SW of the R/P.
    Also in this area are at least six (6) entrances to underground install-
    ations, the extent of which is unknown. The dam of the KUSONG Reservoir
    approx 7000' NW of the R/P, and can supply all needs for water on the
    A/F. Five (5) ammo storage revetments are being built in close prox-
    imity to the R/W and taxiways. There is one (1) at each end of the R/W,
    the other three (3) are in the A/C revetment dispersal area.

10. Buildings: All buildings in close proximity (2 miles) of the A/F area
    are native hut structures. Fifteen (15) of these huts measure approx
    40' X 20' and could be utilized as small work and repair shops. There
    are no conventional A/F type buildings in the A/F area.

11. Radio:There were no radio facilities, other than the radar control for one
    of the heavy gun positions, present on the photography of 18 Oct 51.

12. Defenses: Photography of 18 Oct 51 revealed three (3) heavy eight (8)
    gun batteries and twenty-eight (28) automatic weapons positions. All of
    these gun positions, both heavy and light, are within 2350 yards of the
    R/P. One (1) heavy gun position is controlled by radar. Refer to 67 RTS
    AAA Report #312, dtd 18 Oct 51 for exact locations.

-3-

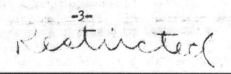

Restricted

13. Transportation: The TEI-SAKU rail line runs from CHONGJU via NAMSI - KUSONG - SAKCHU and crosses the YALU River at CHONGSU-RI four (4) miles West of the SUIHO Dam. An all-weather highway that follows this same route breaks away from the rail line at the YALU River and it runs South-west parallel with the YALU River, crossing into MANCHURIA at CHANG-TIEN-HO-KOU and ANTUNG.

14. Camouflage: None definitely detected at present.

15. Misc Facilities: None noted at present.

16. Expansion Possibilities: Expansion to the North-West is limited to approx 3000' due to low hills and the dam to the KUSONG Reservoir. The R/W can be extended a great distance to the South-East.

17. Aircraft Activity: No A/C have been noted on this A/F to date.

18. Other Activity: The only activity noted has been the construction of the R/W, A/C revetments, supply bunkers, vehicle revetments, and defenses.

19. Remarks: Supplements to this report will be issued as major developments occur. Refer to 67 Rcn Tech Sqdn MRIR Reports for current information on A/C activity and R/W serviceability.

Interpreted By: Approved By:

D. HALL, SGT, USAF

                                         A A GRUMBINE
                                         Major, USAF
                                         OIC, PI Department

DISTRIBUTION:
Special Airfield List

-4-

Restricted

Enclosure recognition photo of Namsi Airfield, used in Special Airfield Report No. 21.

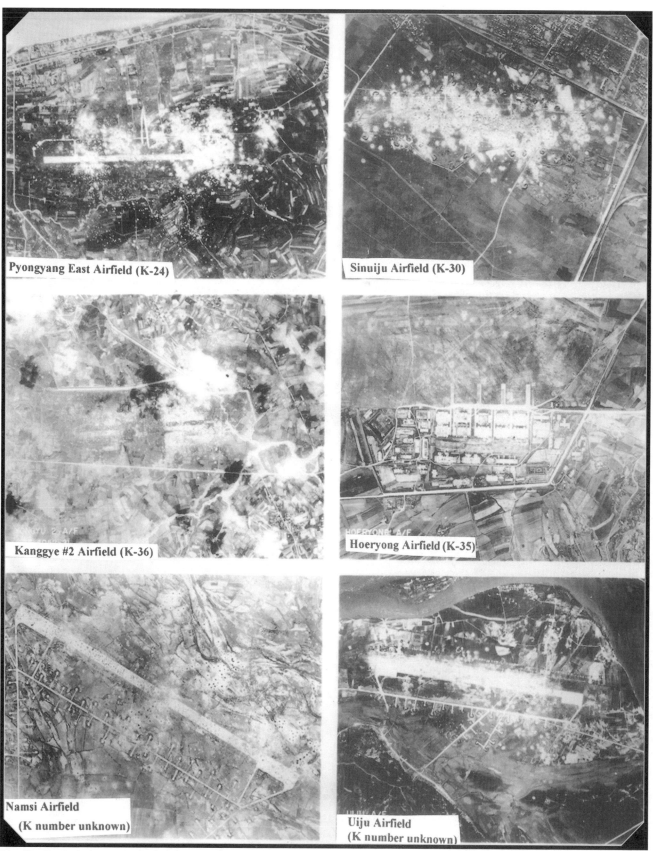

Pyongyang East Airfield (K-24)

Sinuiju Airfield (K-30)

Kanggye #2 Airfield (K-36)

Hoeryong Airfield (K-35)

Namsi Airfield (K number unknown)

Uiju Airfield (K number unknown)

Identification and recognition photos of North Korean airfields.

Captain Lyman Beck. I do not remember the name of the navigator. The photo lab people were there loading film magazines on the cameras. The line crew were loading flash bombs needed for the photos we wanted.

We had a three-man crew: pilot, navigator, and myself. When we boarded the RB-26, the pilot took his left-hand seat, the navigator went to the right (co-pilot) seat, and I was placed in what they called the jump seat, be-

hind and between the pilot and the navigator.

Captain Beck fired up the engines and we taxied out to the end of the runway, where he stopped the plane for the engine checkout. He ran up one engine and let it wind down, then ran up the other and let it wind down; then he ran up both.

We were ready to take my maiden mission into enemy territory in a warplane. What a

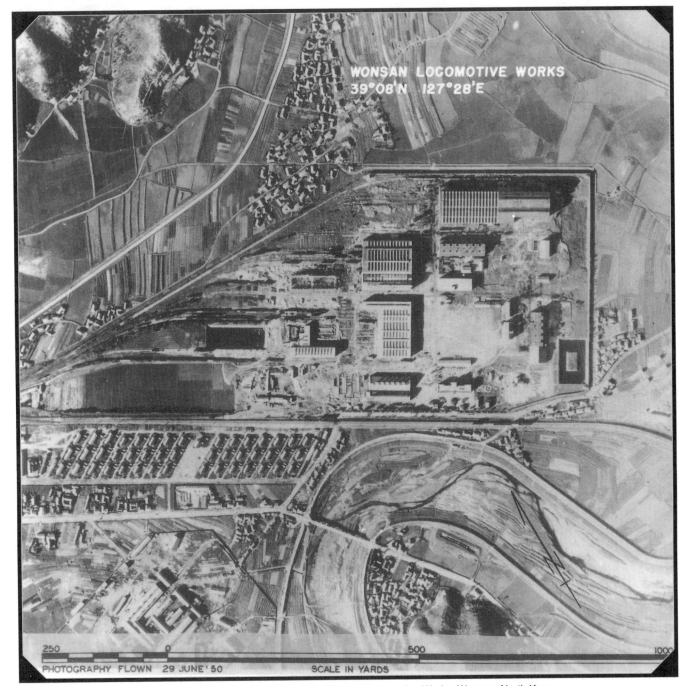

WONSAN LOCOMOTIVE WORKS
39°08'N 127°28'E

250   0   500   1000

PHOTOGRAPHY FLOWN 29 JUNE '50   SCALE IN YARDS

The pre-strike reconnaissance photo, taken June 29, 1950, of the Wonsan Locomotive Works, Wonsan, North Korea.

sensation! As we raced down the runway, I could hear and feel the power of the two engines that seemed to say, *"We'll get you there, and we'll get you back!"* I deeply sensed the true meaning of "off we go into the wild blue yonder." And dark as hell, too!

As we climbed to around 12,000 to 13,000 feet, the navigator moved down into the nose of the RB-26 and I moved into the co-pilot seat he had vacated. I adjusted my parachute and got strapped in. Next I put on the headset so that I could hear what was going on. The sound was completely dead. Within two or three minutes after liftoff from Kimpo Air Base, we were beyond the front lines. I knew in my mind that those fellahs down there didn't like me, but that was OK, as I didn't like them either.

Strike photo of the Wonsan Locomotive Works, August 10, 1950 – a near-perfect bomb pattern. Note one stray bomb cut the rails at the choke point of the marshaling yard. This was saturation bombing at its best, from a B-29 of the 22nd Bombardment Group out of Kadena Air Base, Okinawa.

Low-level nose-oblique aerial photos on this and the facing page show the complete destruction of the Wonsan Locomotive Works from the saturation bombing of August 10, 1950.

After trying three or four times to communicate with Captain Beck with no response, I assumed the order of the day was radio silence. I never did figure out how we got from here to there and back without communication between the pilot and the navigator. Later I would learn.

As we flew deeper into enemy territory on our way to the target, I noted there was sure a lot of cloud cover. I hoped that when we arrived at Hoeryong Airfield we could shoot the photos I thought we were sent to get. About an hour into our flight I began thinking about the Special Airfield Report I had written. I had not noted any antiaircraft positions. I kept telling myself that the reason I hadn't was because there was no enemy artillery — at least I couldn't find any, and I had searched hard. I realized at that moment that this flight would soon be a test to see how thoroughly I did my work.

We did go in and make the full mission, if

The post-strike bomb-damage-assessment reconnaissance photograph of the Wonsan Locomotive Works, Wonsan, North Korea, taken October 10, 1950.

that means anything. Occasionally through the cloud cover we could very briefly see a light on the ground. Because I did not have any radio communication, I didn't have much sense of where we were. I was well acquainted with North Korea through the study of aerial photos, but now, in flight, it was dark and cloudy.

I knew that the U.S. Air Force had pretty well knocked out nearly all of the electric power plants in North Korea, and indeed there were very few lights. Occasionally I would see a solitary glow from possibly a motor vehicle or maybe a campfire. But as soon as our approaching aircraft was heard, the light would be extinguished. This made me feel comfort-

able and confident that the sound of our powerful engines could strike fear into the enemy.

We were quite some time into our flight when I began to think that we should be approaching our target. I believe now, long after the fact, that we then *were* in our target area. However, that area was completely cloud covered. We would get no photos of Hoeryong Airfield that night.

Very soon our entire scene was about to change, and I would learn what it was really like to be flying in a U.S. warplane in an enemy war zone. The clouds began to break up some, and suddenly we saw thousands of lights. I knew instantly where we were, as did Captain Beck, and soon enemy antiaircraft

Months after the bombing of the Wonsan Locomotive Works, reconnaissance photos show no attempt to repair or restore the shop facilities. Many unserviceable freight and passenger cars have been left as is. Photo interpreters wondered if, at the time, the rolling stock could have been carrying personnel and materiel to the front-line area.

fire began appearing in the sky around us. At first it wasn't very close, so we knew the enemy did not have a good fix on us. Captain Beck finally looked over to see how I was handling the situation. From the dim lights off the instrument panel I could see a confident grin on his face. I gave him a "thumb's up," and he knew everything was under control.

Then I heard a loud explosion. I looked out to the side and the sky was as light as day. Immediately I knew we had dropped a flash bomb — *hot damn dog!* We were taking photos! Soon another flash bomb exploded, and we were taking more pictures. I didn't believe this was the place we were supposed to be photographing, however, according to the plan, but I was proud to have been along

An oblique photo, taken on an October 17, 1951, flight east of the Yalu River (foreground), North Korea. Antung Airfield is in the center of the photo; in the far distance, at the left edge of the photo, is Ta-tung-kou Airfield (arrow). The mission was flown by the 15th Tactical Reconnaissance Squadron. Oblique photos like this helped the Americans keep track of Communist forces in Manchuria.

for the ride. Little did those below realize it, but part of the harbor facilities at Vladivostok, Russia, had been photographed in full living black and white from a U.S. Air Force RB-26!

As I watched, tracers and other heavy artillery began to get closer. Captain Beck had stayed the course until he had the photos he wanted, then he began some evasive maneuvers and flew us out over the Sea of Japan, receiving very little damage. I learned not to underestimate the skill and courage of an unarmed reconnaissance pilot.

Out over the sea, we took a southern course until about the Wonsan area. There we were back over land on our return to Kimpo Air Base. When we landed at Kimpo, the photo lab crew was waiting to retrieve the film magazines and take them to the 67th RTS lab for processing. Captain Beck suddenly growled at the Lieutenant navigator for getting us lost and winding up in the wrong country, but I now believe this was merely a smoke screen for my benefit. The line crew asked the Captain, "How was your flight?" The reply: ". . . Just another day at the office."

When we were checking in our parachutes, Captain Beck asked what I thought of the mission. I replied how proud I was to have flown with him, and he answered, "Anytime you want to be a ride-along, just let me know."

After about four hours sleep, I reported to my usual work station. As I went about my duties, an envelope was delivered to me. Inside were two photos and a note. The photos were of a very good quality, especially for night pictures. Normally the print number, mission number, and flying organization would be printed on the photos. All this information was missing. The note read: "Unable to plot, unable to identify, *mission did not happen*." The note was unsigned.

The message and meaning came across loud and clear. I realized why my headset was not plugged in. I was not supposed to know our true target.

I have proudly protected these two treasured photos for more than 50 years. Many times I have asked myself, "How would I have

been described if I had been shot down — KIA, MIA, POW . . . or AWOL?" These ride-along missions were absolutely not authorized.

I believe I would have liked to have been referred to as "Former Staff Sergeant, United States Air Force."

But getting back to the usual operation procedures, an amusing occurrence took place when the Officer in Charge of the Photo Intelligence Department, Major Grumbine, went on R&R to Japan. He had left Lieutenant Hardy in charge of screening and releasing the Mission Review Intelligence Reports in his absence. In one of the reports, the Lieutenant had written that a "clobbered-in aircraft" was at the end of the runway on Uiju Airfield. On Major Grumbine's return, he reviewed the reports that had been released while he had been gone. He challenged the statement about the "clobbered-in aircraft." Lieutenant Hardy asked, "You know what a 'clobbered-in aircraft' is, don't you?" The reply came back, "Yes, but people in the Pentagon might not know what you mean." Hardy's response was that if they don't, they shouldn't be reading that report. The discussion ended as Major Grumbine walked away shaking his head. Needless to say, Lieutenant Hardy was not left in charge of screening and signing reports on Major Grumbine's subsequent R&R trips.

Probably the most frustrating period for the photo interpreters in the 548th Reconnaissance Technical Squadron at Yokota Air Base had been in the fall of 1950. There has been much controversy, still raging, over who said what or who didn't do something on the occurrence of the following. The PIs had been reporting on the masses of Chinese troops that were crossing the Yalu River into western North Korea in the Sinuiju area. Large quantities of equipment and supplies were being stockpiled in that region. But there was no response to their information and the men wondered why nothing was being done to attack this concentration of enemy troops and matériel. More discouraging nearly 50 years later was to read in Jim Mesko's book, *Air War Over Korea*, that "Aerial reconnaissance had not detected the Chinese buildup and the troops were caught totally unaware." Similar remarks can be found in other books and articles.

Although earlier research, done in 1963, proves Mesko's statement to be false, thus far the authors of this work have seen only one instance where that canard has been refuted. The historical record shows that General Douglas MacArthur's staff *was* aware of what we had been reporting, but "sat on" the information and submitted erroneous reports to comply with the General's belief that the Chinese would not get involved in the Korean War. Our troops thus were unaware of the massive Chinese

This photo, showing part of the harbor area of Vladivostok, Russia, is from *"the mission that never happened,"* flown by the 12th Tactical Reconnaissance Squadron, in October 1951, on which Duane Hall was a "ride-along."

buildup, as the information was not passed down to them, where it was so badly needed. Aerial reconnaissance pilots had done their jobs, and the photo interpreters had done theirs. At that time, the PIs at the lower level never were able to fathom why their intelligence information was not responded to by the upper level High Command, and countermeasures taken. They now know some of what had been transpiring at Far East Command Headquarters in the Dai Ichi building in Tokyo.

During the war, one major disadvantage of being

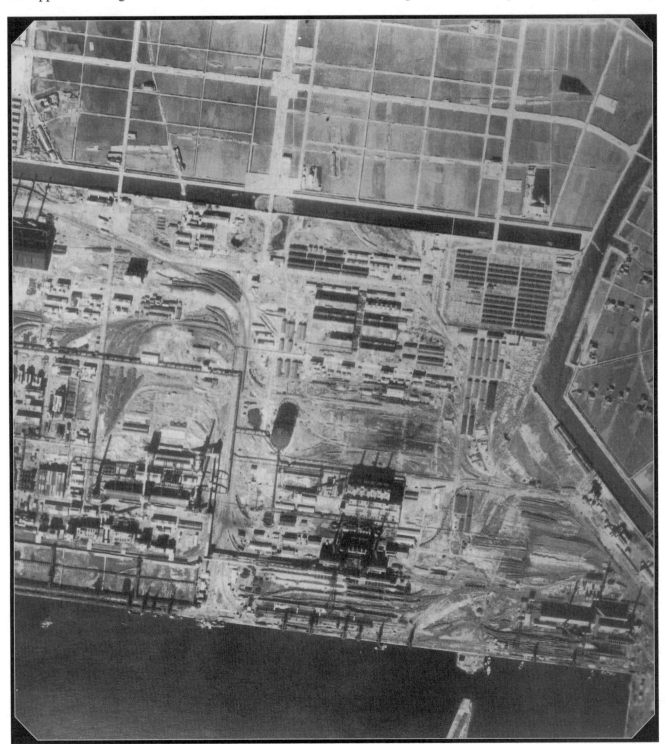

Close-up of the port facilities and industrial area of Vladivostok, Russia, from *"the mission that never happened."*

so close to the front line was the harassment from the previously mentioned North Korean PO-2 aircraft. On nights with a full moon they could usually depend on "Bed Check Charlie" paying a visit. One of the most memorable and least expected visits, as there was *no* full moon, was Christmas Eve 1951. With the temperature way below freezing and a cold wind blowing out of Siberia, the photo interpreters had been partaking of some "anti-freeze" earlier in the evening to ward off the chill. "Charlie" later forced them out of bed and in the bone-piercing cold wind they shivered and cursed in their slit trenches as they futilely fired their carbines at him — with no recorded hits.

As the war progressed, and the Chinese Army became involved, Russian pilots flew their MiG-15s out of air bases in Manchuria. These MiGs proved to be fatal to the RF-80 pilots and the very vulnerable RB-29s. Late in the spring of 1951, the 67th Tactical Reconnaissance Wing Commander, Colonel Karl Polifka, had contacted Wright-Patterson Air Force Base Research and Development Section to investigate the possibility of converting an F-86 into a reconnaissance aircraft. He received a negative reply. He then contacted Far East Air Forces Headquarters in Tokyo and asked their permission to modify one in the field. FEAF took a non-committal stand, letting Colonel Polifka know he was on his own. The Colonel then went up to Kimpo Air Base with some of his photographic support people. The 4th Fighter Interceptor Wing was stationed there with its F-86s. Working with some of the personnel from the 4th FIW on carcasses of damaged and unserviceable aircraft, the Colonel and his men were able to determine that conversion indeed was feasible, with minor interior modifications. The solution was to remove most of the six .50-caliber machine guns in the nose and mount the cameras horizontally; with a mirror set at an angle of 45 degrees, they were able to overcome the space problem.

But before the project could progress to the actual conversion, Colonel Polifka was shot down, on July 1, while flying a reconnaissance mission. Colonels Vincent Howard and Edwin S. Chickering, subsequent Commanders of the 67th TRW, continued to pursue the project. They had an F-86 fighter flown over to Far East Air Materiel Command (FEAMCom) Tachikawa Air Depot in Japan

and had it converted into a reconnaissance aircraft. And thus the birth of the RF-86 — code name "Ash Tray." The original RF-86 had no visible outward modification, which proved to be a very successful answer to the MiG-15 problem in the northwest corner of North Korea in MiG Alley.

The first reconnaissance mission flown by the "Ash Tray" in North Korea was during the winter of 1951-1952. The entire project was classified Top Secret, and the aircraft retained the markings of the 4th FIW. It was parked with the normal F-86s in an attempt to maintain its status undetected by the Communists. However, there was a serious problem in this initial effort, caused by the vibration of the mirrors, which resulted in poor-quality photo images. The problem was solved by mounting the mirrors on the camera installation, rather than directly on the airframe. This led to the code name "Honey Bucket" for this modification. In February 1952, however, ground fire eventually took a toll on the pilot, with minor injuries, and damage to Honey Bucket #1. The pilot was able to bring his aircraft back to Kimpo Air Base for a safe landing, and it was then sent to FEAMCom for repairs. Subsequently, in March, an RF-86 was destroyed in a local aircraft accident caused by fuel starvation.

Eventually six or seven models were produced using the new F-86F version. These models had two downward-pointing, 40-inch, focal-length cameras installed, one on each side of the cockpit compartment. In order to accommodate the film magazines, a "blister" was installed on each side of the cockpit, giving what has been described as a chipmunk effect. A vertical six-inch focal-length camera also was installed as an aid in plotting the missions flown in these new "Haymakers." However, these RF-86Fs were modified *after* the Korean War, and therefore did not fly combat recon missions during the conflict. They did later make "Over-Flights of Denied Territory" — China, the Soviet Union, and North Korea — following the truce in July 1953. As far as we can determine, the missions that were flown by the Haymaker aircraft were in a period roughly between March 1954 through April/May 1956. Captain Hardy subsequently served as one of the photo interpreters on some of those missions.

It was unfortunate that Colonel Polifka never saw his vision fulfilled.

A formation of six "Haymaker" type (code name) RF-86Fs that had been converted into photo reconnaissance aircraft after the Korean War, and thus they never flew combat missions during the conflict.

✈ ✈ ✈ ✈

Occasionally, during the war, Hardy and Hall were fortunate to have a rare day off when the reconnaissance aircraft were grounded due to inclement weather over North Korea. They then took advantage of being *touristas*. "Our adventurous natures," states Hardy, "did not allow us to simply stare at the four walls of our tents during off-duty hours. Knowing this station in Korea was probably a once-in-a-lifetime opportunity, we would leave the air base and hitch rides to various areas. But we always were armed with a .30-caliber carbine, a Colt .45 pistol, and a 35mm camera."

When the pair were at Taegu, they explored and did a little souvenir shopping and some camera

Above and right: The RF-86F Haymaker. Notice the "blister" beneath the pilot's cockpit in the right-hand photo, which was to accommodate the film magazines on the aerial cameras.

work. They were also lucky enough to locate a Korean family outside of town that was willing to do their laundry; and they checked out the large irrigation reservoir east of K-2 (Taegu Air Base), which they had spotted on a set of practice reconnaissance photos. The reservoir made a handy swimming pool to help beat the hot and humid weather.

After the 67th RTS moved to K-14 (Kimpo Air Base), Lieutenant Hardy, Sergeant Hall, along with Sergeants Dave Hitson and Dan Mason, continued to explore the surrounding countryside. Inchon was one of their visits, where they came across three hulls of unfinished Japanese miniature submarines from World War II. Other trips took them into Seoul, the South Korean capital, where they found the ravished capitol building, which they entered by a side door. The building has since been razed; the new capitol building has been erected on the site of the former Seoul Airfield (K-16) on an island in the Han River. Ben Hardy recalls, "When we exited the front entrance of the capitol building that day we encountered a huge sign announcing, 'DO NOT ENTER BUILDING IS MINED.'"

Continuing their status as *touristas*, the pair posed in front of the old original South Gate (now listed as "Korean Treasure Number 1"), at that time the entry into Seoul. Another stop was at the famed Seoul Triple Railroad Bridges that had picked up the nickname "The Plastic Bridges" from the difficulty that Bomber Command had in knocking them down. A particularly rewarding experience they enjoyed was related to their sparse off-duty activities. Four of the men (Hall, Hitson, Taylor, and Hardy) sponsored a Seoul orphanage. When they made their R&R trips to Japan, they always managed to find some useful items for the children. They often were needled about the full duffel bags they carried on their return trips to Kimpo.

Late in the spring of 1952, the men who had transferred to the 67th Reconnaissance Technical Squadron from the 548th RTS started their rotation back to the States (sarcastically referred to by GIs as the "Land of Technicolor Staff Cars" — a comment on the military's olive-drab sedan automobiles).

Both the 548th RTS and the 67th Tactical Reconnaissance Wing, and various reconnaissance (recce)

outfits, subsequently have had several reunions where old hands enjoy a few days of friendship and reminiscing about the "old days." Hardy notes, "We all believe that the activities we participated in while stationed in Japan and Korea as a support group to the ground forces made a difference. Over the many years since then, some of us still enjoy occasional telephone or letter contact."

In retrospect, Colonel Karl Polifka, Commander of the 67th Tactical Reconnaissance Wing, had obviously assembled and trained a very capable team of personnel. Many of his key officers and NCOs were veterans of World War II. Even down at the 67th Reconnaissance Technical Squadron level, many of the officers and NCOs were proven veterans; some had served in the Army Air Forces and others in the Navy, joining the U.S. Air Force in 1947, when it became an entity in its own right. Some of his pilots also had flown during World War II and continued to maintain their status on reconnaissance missions in Korea.

When Colonel Polifka was shot down in July 1951, he had been flying an RF-51 over the front line in Korea. He was a large man and apparently had elected to bail out of his stricken aircraft at low altitude. His parachute streamed out and became entangled with the tail of his plane, dragging him down. But the Colonel died doing what he liked best. The next day at noon, many of the men at Taegu Air Base attended his memorial service, and none will ever forget "The Missing Man Fly-Over" for the Colonel: Major Fish flew an RB-26, Major Woodyard flew an RF-80, and Lieutenant Carroll flew an RF-51. Colonel Karl Polifka had commanded a very loyal and proud group.

On June 4, 1951, the 548th Reconnaissance Technical Squadron was awarded the Meritorious Unit Commendation for service from June 27, 1950, to April 10, 1951. On April 7, 1955, the 548th RTS was awarded the Air Force Outstanding Unit Award for service from April 11, 1951, through November 27, 1954.

The 67th Tactical Reconnaissance Wing was awarded the Air Force Outstanding Unit Award for service from December 1, 1952, to April 20, 1953.

Lieutenant Ben Hardy (left) and Sergeant Duane Hall (below) at the Seoul Triple Railroad Bridges — The Plastic Bridges — fall 1951.

Sergeant Dan Mason (left) and Sergeant Dave Hitson (below) at the Seoul Triple Railroad Bridges, fall 1951.

◄ Sergeant Dan Mason, Lieutenant Ben Hardy, and Sergeant Duane Hall on one of the bomb-destroyed Seoul Triple Railroad Bridges, fall 1951.

The 67th Tactical Reconnaissance Group was awarded three Presidential Unit Citations for service from February 25, 1951, to April 21, 1951; July 9, 1951, to November 27, 1951; and May 1, 1952, to July 27, 1953.

For many years, some of the men of these units

have been members of the Korean War Veterans Association (The Greybeards) and the Veterans of Foreign Wars (VFW), looking forward to the publications of those organizations as well as *The Recce Reader*, *The Commemorator*, and *Military* magazine. The men anxiously anticipate articles that tell

The three railroad bridges in the upper half of this photo earned the nickname "The Plastic Bridges" from Bomber Command's difficulty in trying to knock them down, even after multiple attacks. The larger, heavily constructed bridge had a plate attached to one of the girders that read "Built by the American Bridge Company 1911." Note that there is one span missing in each of the other bridges. After a combined effort of U.S. Navy Corsairs and Skyraiders and U.S. Air Force B-29s, the problem was taken care of August 19-20, 1950. (Oblique photo by 15th Tactical Reconnaissance Squadron, July 16, 1950.)

OL 15TRS R2805A 5AF 4MAR51

Six months after the *coup de grâce*, on March 4, 1951, "The Plastic Bridges" remained unserviceable. (Left oblique photo, flown by 15th Tactical Reconnaissance Squadron.)

the story of the American ground force individual or units. They remain gratified by the knowledge that their photo intelligence/interpreter contribution made most of those stories close with a successful ending.

✈ ✈ ✈ ✈

Not too long after the truce was signed at Pan-

munjom on July 27, 1953, uncertainty and chaos returned to the tactical reconnaissance capability.

In summary, the 67th Tactical Reconnaissance Wing moved to Itami Air Base, Japan, in late 1954. On July 1, 1957, it moved to Yokota Air Base, Japan, and assumed all reconnaissance requirements for Far East Air Forces. On December 8, 1960, it was deactivated, then reactivated on August 2, 1965, at Mountain Home Air Force Base, Idaho.

Bombed-out buildings at Inchon, South Korea, fall 1951.

Sergeant Dave Hitson (left), Lieutenant Ben Hardy (center), and Sergeant Duane Hall at the original entry gate to Seoul, fall 1951.

The capitol building in Seoul, South Korea, summer 1951. The building was completely gutted by fire. After the war it was razed and a new capitol building was erected on the site of the former Seoul Airfield, on an island in the Han River.

Below: The 1951 interior view of a stripped World War II Japanese miniature submarine fabrication and repair facility at Inchon, a remnant of the Japanese occupation of Korea.

An unfinished miniature World War II Japanese submarine, adjacent to the fabrication and repair facility at the Inchon Harbor area, 1951. Note the Japanese flag on the conning tower.

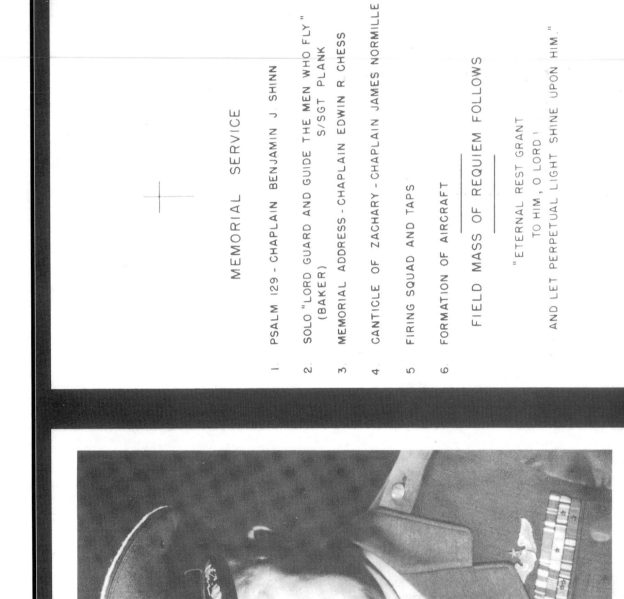

MEMORIAL SERVICE

1. PSALM 129 - CHAPLAIN BENJAMIN J SHINN

2. SOLO "LORD GUARD AND GUIDE THE MEN WHO FLY"
   S/SGT PLANK
   (BAKER)

3. MEMORIAL ADDRESS - CHAPLAIN EDWIN R. CHESS

4. CANTICLE OF ZACHARY - CHAPLAIN JAMES NORMILLE

5. FIRING SQUAD AND TAPS

6. FORMATION OF AIRCRAFT

FIELD MASS OF REQUIEM FOLLOWS

"ETERNAL REST GRANT
TO HIM, O LORD !
AND LET PERPETUAL LIGHT SHINE UPON HIM."

The program from Colonel Karl L. Polifka's memorial service. Colonel Polifka, Commander of the 67th Tactical Reconnaissance Wing, was shot down over Korea on July 1, 1951.

The 67th TRW reorganized on January 1, 1966. On July 15, 1971, the 67th moved to Bergstrom Air Force Base, Texas. It was redesignated the 67th Reconnaissance Wing on October 1, 1991. The 67th Reconnaissance Wing was deactivated September 30, 1993, and replaced on October 1, 1993, by the 67th Intelligence Wing; it is now stationed at the KELLY USA business park — formerly Kelly Air Force Base, which was transferred to the city of San Antonio and renamed. The new units stationed there are now under the jurisdiction of Lackland Air Force Base, Texas, for logistical support.

The 67th Tactical Reconnaissance Group moved to Itami Air Base, Japan, and was reduced to paper status on June 1, 1954, then in late 1954 was manned again. In August 1956, the 67th TRG moved to Yokota Air Base, Japan.

The 67th Air Base Group and components moved to Itami Air Base, Japan; on July 1, 1957, it was inactivated.

The 67th Maintenance and Supply Group and components moved to Itami Air Base, Japan; on July 1, 1957, it was inactivated.

The 67th Tactical Hospital moved to Itami Air Base, Japan; on July 1, 1957, it was inactivated.

The 67th Tactical Reconnaissance Wing Detachment 1 remained at Kimpo Air Base, Korea, with its RB-26 Invader aircraft.

The 67th Reconnaissance Technical Squadron moved to Itami Air Base, Japan, on July 1, 1957, eventually moving to Yokota Air Base, Japan, replacing the 548th Reconnaissance Technical Squadron.

The 11th Tactical Reconnaissance Squadron moved to Itami Air Base, Japan, with its RB/WB-26 Invader photo reconnaissance/weather aircraft. In September 1966 it moved to Southeast Asia.

The 12th Tactical Reconnaissance Squadron moved to Itami Air Base, Japan, on November 8, 1954. On August 14, 1956, it moved to Yokota Air Base, Japan, where it was equipped with Douglas RB-66 Destroyer tactical bomber/photo recon aircraft. In April 1958, two of the 12th's RB-66s supported a Southeast Asia Treaty Organization (SEATO) exercise. On March 8, 1960, the 12th TRS was deactivated at Yokota Air Base, Japan. On July 1, 1966, it was reactivated at Mountain Home Air Force Base, Idaho. On September 2, 1966, the 12th TRS moved to Tan Son Nhut Air Base, Republic of Vietnam. On August 20, 1971, it moved to Bergstrom Air Force Base, Texas. It was later deployed to Bahrain from January 10 to May 12, 1991. The 12th TRS ended flying operations in August 1992. It is currently (2003) stationed at Beale Air Force Base, California, operating the Global Hawk unmanned aerial vehicle.

The 15th Tactical Reconnaissance Squadron returned to Komaki Air Base, Japan, in March 1954. In August 1955 the squadron moved to Yokota Air Base, Japan, and converted to the North American RF-86 Sabre Jet. Subsequently, while there, the 15th converted to the Republic RF-84 Thunderjet. In August 1956 the 15th TRS moved to Kadena Air Base, Okinawa.

The 45th Tactical Reconnaissance Squadron moved to Misawa Air Base, Japan, in March 1955 and converted to RF-84Fs. In July 1971 the squadron moved to Bergstom Air Force Base, Texas. It operated out of England in June and July 1973, then was inactivated in October 1975.

The 91st Strategic Reconnaissance Squadron became the 91st Tactical Reconnaissance Squadron and was assigned to the 67th Tactical Reconnaissance Wing at Bergstrom Air Force Base, Texas. The 91st operated out of Italy and Greece in April and May 1972.

The 548th Reconnaissance Technical Squadron was replaced by the 67th Reconnaissance Technical Squadron and moved to Hickham Field Air Force Base, Hawaii.

✈ ✈ ✈ ✈

HEADQUARTERS
FAR EAST AIR FORCES
APO 925

GENERAL ORDERS)
             :
NUMBER   254)

MERITORIOUS UNIT COMMENDATION

By direction of the Secretary of the Air Force, under the provisions of Air Force Regulation 35-75 and pursuant to authority contained in Section V, General Orders Number 3, Department of the Air Force, 23 January 1951, the Meritorious Unit Commendation is awarded to the following named unit of the United States Air Force, for exceptionally meritorious conduct in the performance of outstanding services during the period indicated. The citation reads as follows:

The 548th Reconnaissance Technical Squadron distinguished itself by exceptionally meritorious conduct in performance of outstanding service from 27 June 1950 to 10 April 1951. Throughout this period, the 548th Reconnaissance Technical Squadron, despite a shortage of space, adequate equipment and experienced personnel, produced voluminous quantities of superior photographic, interpretative, and reproduction material. During the month of August 1950, for example, over two hundred thousand photographic prints and two and a half million impressions by the printing plant were processed. The work accomplished by the 548th Reconnaissance Technical Squadron was of invaluable aid to the United Nations forces in Korea. Countless hours of overtime were willingly devoted by personnel of the Squadron to the arduous task of supplying strike photographs, target materials and allied photographic requirements of Far East Air Forces units. The speed and skill with which the organization processed reconnaissance film greatly aided in the planning for and the successful accomplishment of the Inchon landing. In addition to these outstanding contributions to the United Nations efforts in the Korean conflict, the 548th Reconnaissance Technical Squadron fulfilled numerous and varied requests for units of the Allied Occupation Forces in Japan, and from the United States Air Force in the Zone of the Interior. Outstanding among these jobs was the accomplishment over a six months period of many complicated assignments for Project "Wringer," and the Special Photo Intelligence Reports furnished Headquarters, United States Air Force. The latter reports were of great value in the production of further intelligence and resulted in a request by Strategic Air Command for inclusion on future distribution lists. Regardless of the pressure under which it was necessarily produced, the quality of work of the 548th Reconnaissance Technical Squadron was consistently superior and all productions were completed on schedule. This outstanding accomplishment, which frequently required many extra hours of exhausting work, was possible only through the complete cooperation and unremitting efforts of every member of the organization. The meritorious conduct and devotion to duty displayed by the personnel of the 548th Reconnaissance Technical Squadron were of immeasurable assistance to the United Nations forces in Korea, the Allied Occupation Forces in Japan, and the United States Air Force in the Zone of the Interior in the accomplishment of their respective missions and reflected great credit upon the Squadron, the Far East Air Forces, and the United States Air Force.

GO No. 254, Hq FEAF, APO 925, 4 June 51, cont'd

BY COMMAND OF LIEUTENANT GENERAL PARTRIDGE:

OFFICIAL:

                                         L. C. CRAIGIE
                                         Major General, USAF
                                         Vice Commander

E. E. TORO
Colonel, USAF
Adjutant General

1st Ind

HEADQUARTERS, 548th Reconnaissance Technical Squadron, APO 328, 11 June 1951

TO: M/Sgt Benson B. Hardy   AF 17227171

    1. You are permanently authorized the award of the Meritorious Unit Commendation.

    2. <u>Meritorious Unit Commendation</u>:

        a. <u>Description</u>. The Meritorious Unit Commendation awarded during World War II consists of a scarlet streamer embroidered in white with the name of the theater or area of operations in which the service was rendered. Each award of the Meritorious Unit Commendation is marked by a separate streamer. The individual emblem for the Commendation consists of a gold colored laurel wreath embroidered on a small square of cloth matching the uniform material. Gold colored numerals denote repetitive awards. Number two denotes the first repetition.

        b. <u>When Authorized</u>. The Meritorious Unit Commendation, established by the War Department, was awarded to units of the Armed Forces of the United States or its allies for exceptionally meritorious conduct in the performance of outstanding services for at least six months during the period of operations against an armed enemy of the United States. The unit must have displayed such outstanding devotion or superior performance of exceptionally difficult tasks that it set the unit apart and above other units with similar missions.

        c. <u>Who May Wear</u>. Personnel assigned or permanently attached to and present for duty at least 60 days with a unit during the period for which the Commendation was awarded are eligible for permanent award of the emblem. Personnel subsequently assigned or permanently attached to a unit which has received a citation but who were not present with the unit during the cited period may wear the emblem only for the duration of assignment or attachment. <u>The individual emblem for the Meritorious Unit Commendation will not be worn on the Air Force uniform</u>.

                                              GEORGE H FISHER
                                              Major, USAF
                                              Commanding

Afterword

The purpose of this book has been to preserve some of the documents, photos, and memories of a Lieutenant and a Staff Sergeant in photo intelligence/photo interpreter work and to relate a few of their experiences during their tour of duty during the Korean War. Most, if not all, of the information or material used, in general, is readily available. Although similar material may exist in archives, most of the photos and reports used herein have a personal connection to the authors.

Relying on 50-year-old memories and often scant published materials, there is a potential for error, as in a mix-up in the different code names assigned to the various changes and modifications of the RF-86. Should such be evident in this work, the authors apologize. We might note that an RF-86 (52-4492) was on display at Bergstrom Air Force Base, Texas, until the base was closed in September 1993. The aircraft was moved to the St. Louis Aviation Museum at Chesterfield, Missouri, then in May 1995, when the Mississippi River flooded, it was nearly destroyed. The plane was rescued in a very sorry condition and is presently being restored at the USAF Museum at Wright-Patterson Air Force Base, Dayton, Ohio.

Photo interpreters have never been officially recognized publicly, but they in fact have contributed massive volumes of intelligence information to many of the campaigns launched in Korea. Most notable were the 548th Reconnaissance Technical Squadron Intelligence Reports and aerial photos provided to Commanding Generals, including General of the Army Douglas MacArthur. These reports helped to make the Inchon landing and the Suk-chon/Sunchon paratroop campaigns the success that they were.

Lieutenant Hardy and Sergeant Hall's assignment at Taegu Air Base, establishing an airfield section within the Photo Interpretation Department, was an "awesome" responsibility, involving the identification and constant monitoring of all enemy activity on all the airfields in North Korea and surrounding areas. As a ride-along passenger on one particular night photo-reconnaissance mission, Duane Hall states that the thought came to mind, "I hope to hell I didn't miss a radar-controlled antiaircraft installation." But they were never shot down, if that means anything,

The Photo Interpretation Departments of the 548th and 67th Reconnaissance Technical Squadrons were very well respected organizations for the accurate and in-depth intelligence information they provided to all branches of the U.S. military as well as to members of other United Nations military participating in the Korean War.

Playing a minuscule role in the 67th Reconnaissance Technical Squadron's ascent to a competent and highly respected focal point — the *raison d'être* of the 67th Tactical Reconnaissance Wing — gave all the photo interpreters a feeling of a "job well done." Serving the U.S. Air Force as photo interpreters, Hardy adds, has been a source of pride, to be able to say, *"We were the first to know."*

And we can't help but note that the Nattering Nabobs (thanks Spiro) of Change have continued their one-upmanship: The decades old Veterans Administration is now The Department of Veterans Affairs and we former photo interpreters have become "images analysts."

Bibliography

Anderson, Jack. "What Really Happened in Korea," *Parade*, September 22, 1963.

Babington-Smith, Constance. *Air Spy*. New York: Harper and Brothers, 1956.

Blair, Clay. *The Forgotten War*. New York: Times Books, 1987.

Brady, James. *The Coldest War*. New York: Orion Books, 1990.

Dorr, Robert F., and Warren Thompson. *The Korean Air War*. New York: Time-Life Books, 1987.

Edwards, Paul M. *The Korean War*. Melbourne, FL: Krieger Publishing Co., 1999.

Endicott, Judy G. *The USAF in Korea. Campaigns, Units, and Stations, 1950-1953*. Washington, D.C.: Air Force History and Museums Program, U.S. Superintendent of Documents, 2001.

FEAF Bomber Command and the Air War in Korea, 1950-1953. Washington, D.C.: Air Force History and Museums Program, U.S. Superintendent of Documents, 2000.

Forty, George. *At War in Korea*. New York: Bonanza Books, 1985.

Foster, Colonel Frank, and Lawrence Borts. *A Complete Guide to All United States Military Medals 1939 to Present*. Fountain Inn, SC: Medals of America Press, 2000.

Futrell, Robert Frank. *The United States Air Force in Korea, 1950-1953*. New York: Duell, Sloan and Pearce, 1961. Rev. ed., Washington, D.C.: Office of Air Force History, 1983, 1991.

Giangreco, D. M. *War in Korea*. Novato, CA: Presidio Press, 1990.

"Heroes," *Time*, April 10, 1964: 24-25; April 17, 1964: 40-41.

Higgins, Marguerite. *War in Korea*. New York: Doubleday and Company, 1951.

Jackson, Robert. *Air War Korea, 1950-1953*. Shrewsbury, UK: Airlife Publications, Ltd., 1998.

Lashmar, Paul. *Spy Flights of the Cold War*. Glou-

cestershire, UK: Sutton Publishing, 1996.

Leary, William M. *Anything, Anywhere, Anytime. Combat Cargo in the Korean War*. Washington, D.C.: Air Force History and Museums Program, U.S. Superintendent of Documents, 2000.

Life, "Speaking of Pictures," July 9, 1951: 6-7; "Inside MiG Alley," December 10, 1951: 30-31.

Mesko, Jim. *Air War Over Korea*. Carrollton, TX: Squadron/Signal Publications, 2000.

Photographic Interpretation Handbook. Washington, D.C.: U.S. Government Printing Office, 1954.

Shaw, Frederick J., Jr., and Timothy Warnock. *The Cold War and Beyond. Chronology of the United States Air Force, 1947-1997*. Washington, D.C.: Air Force History and Museums Pro-gram, U.S. Superintendent of Documents, 1997.

Stanton, Shelby L. *America's Tenth Legion*. Novato, CA: Presidio Press, 1989.

Summers, Harry G. *Korean War Almanac*. New York: Facts on File, Inc., 1990.

The Civil War. New York: Time-Life Books, 1987.

Thompson, Wayne, and Bernard C. Nalty. *Within Limits, The U.S. Air Force and the Korean War*. Washington, D.C.: Air Force History and Museums Program, U.S. Superintendent of Documents, 1996.

Warnock, A. Timothy. *The USAF in Korea. A Chronology, 1950-1953*. Washington, D.C.: Air Force History and Museums Program, U.S. Superintendent of Documents, 2000.

Weintraub, Stanley. *MacArthur's War*. New York: The Free Press, 2000.

Wilcoxen, Kathryn A. *Significant Air Mobility Events of the Korean War*. Scott Air Force Base, IL: Air Mobility Command, 2000.

Y'Blood, William T. *Down in the Weeds: Close Air Support in Korea*. Washington, D.C.: Air Force History and Museums Program, U.S. Superintendent of Documents, 2002.

✈ ✈ ✈ ✈

Appendix A
Known Awards
June 27, 1950, through February 1952

548th Reconnaissance Technical Squadron

Meritorious Unit Commendation	June 27, 1950, to April 10, 1951
Air Force Outstanding Unit Award	April 11, 1951, to November 26, 1954
U.S. Korean Service Medal	June 27, 1950, to July 27, 1953
UN Service Medal with Korea Bar	June 27, 1950, to July 27, 1953
National Defense Service Medal	June 27, 1950, to July 27, 1953
Army of Occupation Medal	1945 to 1955
Republic of Korea Service Medal	June 27, 1950, to July 27, 1953

67th Tactical Reconnaissance Wing

Air Force Outstanding Unit Award	December 1, 1952, to April 30, 1953
Korean Presidential Unit Citation	February 25, 1951, to March 31, 1953
U.S. Korean Service Medal	February 25, 1951, to July 27, 1953
UN Service Medal with Korea Bar	February 25, 1951, to July 27, 1953
National Defense Service Medal	February 25, 1951, to July 27, 1953
Republic of Korea Service Medal	February 25, 1951, to July 27, 1953

67th Tactical Reconnaissance Group

Presidential Unit Citation (3)	February 25, 1951, to April 21, 1951
(Originally, Distinguished Unit Citation)	July 9, 1951, to November 27, 1951
	May 1, 1952, to July 27, 1953
Air Force Outstanding Unit Award	December 1, 1952, to April 30, 1953

Campaign Stars for the U.S. Korean Service Medal

UN Defensive	June 27, 1950, to September 15, 1950
UN Offensive	September 16, 1950, to November 2, 1950
Chinese Communist Forces Intervention	November 3, 1950, to January 24, 1951
First UN Counteroffensive	January 25, 1951, to April 21, 1951

Campaign Stars for the U.S. Korean Service Medal *(continued)*

Chinese Communist Forces Spring Offensive	April 22, 1951, to July 8, 1951
UN Summer-Fall Offensive	July 9, 1951, to November 27, 1951
Second Korean Winter	November 28, 1951, to April 30, 1952
Korea, Summer-Fall	May 1, 1952, to November 30, 1952
Third Korean Winter	December 1, 1952, to April 30, 1953
Korea, Summer	May 1, 1953, to July 27, 1953

Note: This list is undoubtedly incomplete; it mostly covers the period during which the authors served in Japan and Korea. Though the dates of eligibility of the above awards and medals does extend beyond our service there, it does not indicate the entire period of eligibility. Recently there seems to be a plethora of ribbons and medals issued that cover some of the periods above. We have made no attempt to verify or determine eligibility for the above, or for the below listed recently issued items.

Recently Issued Medals:

50th Anniversary Korean Defense, 1950-2003

Cold War Victory Commemorative, 1945-1991

USAF Service Commemorative, 1947-2003

Appendix B
Acronyms and Abbreviations

Many of the following acronyms and abbreviations were commonly used in our photo intelligence work in the U.S. Air Force. Some have appeared within the text; some have been included herein for reader interest.

AB	Air base, on non-U.S. territory
A/C	Aircraft
AFB	USAF base, on U.S. territory
Ammo	Ammunition
AMS	Army Map Service
AWOL	Absent without leave
EO	Executive Order
EPRD	Engineer Photography Reproduction and Distribution
FEAF	Far East Air Forces
FEAMCom	Far East Air Materiel Command
FEC	Far East Command
FIGMO	"Finally I Got My Orders"
FIS	Fighter Interceptor Squadron
FIW	Fighter Interceptor Wing
GI	Originally, Government (or General) Issue, as in "GI boots"; during World War II, a serviceman
IPIR	Immediate Photo Intelligence Report
JOC	Joint Operations Center
KIA	Killed in action
KMAG	Korean Military Advisory Group
MG	Machine gun
MIA	Missing in action
MiG	Russian fighter aircraft named for its designers, Mikoyan and Gurevich
MRIR	Mission Review Intelligence Report
NCO	Non-commissioned officer
NKPA	North Korean People's Army
OIC	Officer in Charge
OL	Oblique left; oblique camera on left side of aircraft
ON	Oblique nose; oblique camera in nose of aircraft
PI	Photo interpreter, photo intelligence
PIO	Public Information Office
POL	Petroleum, oil, lubricants
POW	Prisoner of war
RAF	Royal Air Force (British)
R&R	Rest and recreation (usually three days)
RB	Reconnaissance bomber
RF	Reconnaissance fighter
ROK	Republic of Korea (South)
R/P	Reference point
RTS	Reconnaissance Technical Squadron
R/W	Runway
SAC	Strategic Air Command
SEATO	Southeast Asia Treaty Organization
TDY	Temporary duty
TRG	Tactical Reconnaissance Group
TRS	Tactical Reconnaissance Squadron
TRW	Tactical Reconnaissance Wing
TSG	Tactical Support Group
UAV	Unmanned aerial vehicle
U/I	Unidentified
UN	United Nations
USAF	United States Air Force
USMC	United States Marine Corps
USN	United States Navy
USSR	Union of Soviet Socialist Republics (Russia)
VIP	Very important person
VV	Vertical photography

✈ ✈ ✈ ✈

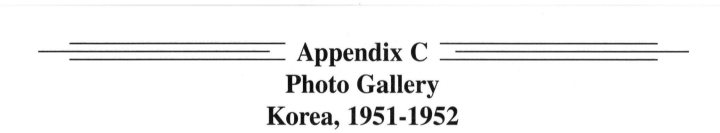

Appendix C
Photo Gallery
Korea, 1951-1952

Right and below: Royal Australian Air Force MK-8 Meteors at Kimpo Air Base, fall 1951. 67th Tactical Reconnaissance Wing aircraft followed the Meteors on airstrikes, to gather damage-assessment photos.

Above and left: The RAAF MK-9 Meteor two-seat trainer at Kimpo Air Base, fall 1951.

A Douglas C-54 Skymaster four-engine transport takes off from Kimpo Air Base, fall 1951. In June 1950, a C-54 was the first U.S. Air Force plane to be destroyed in the Korean War when it was strafed on the ground at Kimpo.

Right: The C-54 Skymaster, parked at Kimpo Air Base, fall 1951.

Above and right: A Lockheed C-121 Constellation, formerly General Douglas MacArthur's aircraft *Bataan,* on the ground at Kimpo Air Base, fall 1951. The aircraft was a four-engine, triple tail-fin cargo transport that was used for electronic surveillance in Korea and Vietnam.

Local Koreans help to repair and enlarge the airfield at Kimpo Air Base, 1951.

The Han River bridges at Seoul, South Korea, in operation after repair by the U.S. Army 62nd Engineer Construction Battalion, 1951.

South Gate in Seoul, South Korea. Upper photo in 1951, lower photo in 2002 during renovation, and now listed as Korean Treasure Number 1.

South Korean children and local farmer carrying his day's work on his back.

Left: Lieutenant Ben Hardy taking a letter-writing time-out at Kimpo Air Base, summer 1951.

Off-duty hours allowed for enthusiastic relaxation. Above is Duane Hall and Ben Hardy's "Membership" certificate into the "International Order of Mutual Admiration Societies."

The men of the 67th Reconnaissance Technical Squadron enjoyed a steak fry at Kimpo Air Base, 1951.

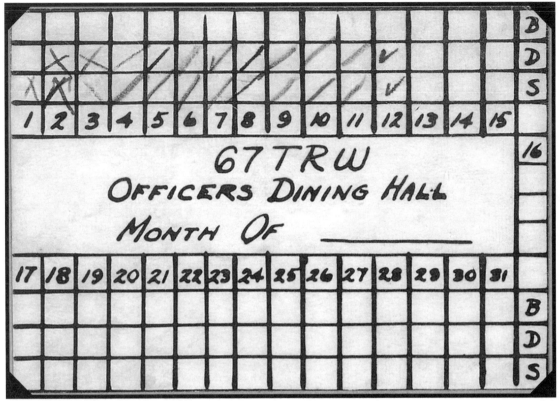

The Officers Dining Hall pass issued by the 67th Tactical Reconnaissance Wing at Kimpo Air Base, 1951.

A Korean "1000 Won" bill from 1951. Today, one Korean Won is approximately equivalent to .0008726 in U.S. dollars.

Index

by Lori L. Daniel

✈ ✈ ✈ ✈